One, Two, Three

One, Two, Three

ABSOLUTELY ELEMENTARY MATHEMATICS

David Berlinski

Pantheon Books · New York

Harem Scene with Sultan, by Jean-Baptiste van Mour, reproduced
courtesy of Azize Taylan and with the assistance of Okan Altiparmak.

Library of Congress Cataloging-in-Publication Data
Berlinski, David, [date]
One, two, three : absolutely elementary mathematics / David Berlinski.
p. cm.
Includes index.
ISBN 978-0-375-42333-8
1. Symmetry (Mathematics) I. Title.
QA174.7.S96B47 2011 510—dc22 2010038555

www.pantheonbooks.com

Jacket design by Brian Barth

Printed in the United States of America

First Edition

2 4 6 8 9 7 5 3 1

For Neal Kozodoy

We read to find out what we already know.

—V. S. Naipaul

One, Two, Three

INTRODUCTION

This is a little book about absolutely elementary mathematics (**AEM**); and so a book about the natural numbers, zero, the negative numbers, and the fractions. It is neither a textbook, a treatise, nor a trot. I should like to think that this book acts as an anchor to my other books about mathematics.

Mathematicians have always imagined that mathematics is rather like a city, one whose skyline is dominated by three great towers, the state ministries of a powerful intellectual culture — our own, as it happens. They are, these great buildings, devoted to Geometry, Analysis, and Algebra: the study of space, the study of time, and the study of symbols and structures.

Imposing as Babylonian ziggurats, these buildings convey a sacred air.

The common ground on which they rest is sacred too, made sacred by the scuffle of human feet.

This is the domain of absolutely elementary mathematics.

Many parts of mathematics glitter alluringly. They are exotic. Elementary mathematics, on the other hand, evokes the very stuff of life: paying bills, marking birthdays, dividing debts, cutting bread, and measuring distances. It is earthy. Were textbooks to disappear tomorrow, and with them the treasures that they contain, it would take centuries to rediscover the calculus, but only days to recover our debts, and with our debts, the numbers that express them.

Elementary mathematics as it is often taught and sometimes used requires an immersion into messiness. Patience is demanded, pleasure deferred. Decimal points seem to wander, negative numbers become positive, and fractions stand suddenly on their heads.

And what *is* three-fourths divided by seven-eights?

The electronic calculator has allowed almost everyone to treat questions such as this with an insouciant indifference. Quick, accurate, and cheap, it does better what one hundred years ago men and women struggled to do well. The sense that in elementary mathematics things are familiar—half remembered, even if half forgotten—is comforting, and so are the calculator and the computer, faithful almost to a fault, but the imperatives of memory and technology do prompt an obvious question: why bother to learn what we already know or at least thought we knew?

The question embodies a confusion. The techniques of elementary mathematics are one thing, but their *explanations* are quite another. Everyone can add two simple natural numbers together—two and two, for instance. It is much harder to say what addition means and why it is justified. Mathematics explains the meaning and provides the justification. The theories that result demand the same combination of art and sophistication that is characteristic of any great intellectual endeavor.

It could so easily have been otherwise. Elementary mathematics, although pressing in its urgency, might have refused to cohere in its theory, so that, when laid out, it resembled a map in which roads diverged for no good reason or ended in a hopeless jumble. But the theory by which elementary mathematics is explained and its techniques are justified is intellectually coherent. It is powerful. It makes sense. It is never counter-intuitive. And so it is appropriate to its subject. If when it comes to the simplest of mathematical operations—addition again—there remains something that we do not understand, that is only because there is nothing in nature (or in life) that we understand as completely as we might wish.

Nonetheless, the theory that results is radical. Do not doubt it. The staples of childhood education are gone in the night. One idea is left, and so one idea predominates: *that the calculations and concepts of absolutely elementary mathematics are controlled by the single act of counting by one.* There is in this analysis an economy of effect, and a reduction of experience to its essentials, as dramatic as anything found in the physical sciences.

Until the end of the nineteenth century, this was not understood. A century later, it is still not *widely* understood. School instruction is of little help. "Please forget what you have learned in school," the German mathematician Edmund Landau wrote in his book *Foundations of Analysis*; "you haven't learned it."

From time to time, I am going to ask that readers do some forgetting all their own.

A secret must now be imparted. It is one familiar to anyone writing about (or teaching) mathematics: no one very much likes the subject. It is best to say this at once. Like chess, mathematics has the power to command obsession but not often affection.

Why should this be—the distaste for mathematics, I mean?

There are two obvious reasons. Mathematics confronts the beginner with an aura of strangeness, one roughly in proportion to its use of arcane symbols. There is something about mathematical symbolism, a kind of peevishness, that while it demands patience, seems hardly to promise pleasure.

Why bother?

If the symbolic apparatus of mathematics is one impediment to its appreciation, the arguments that it makes are another. Mathematics is a matter of proof, or it is nothing. But certainty does not come cheap. There is often a remarkable level of detail in even a simple mathematical argument, and, what is worse, a maddening difference between the complicated structure of a proof and the simple and obvious thing it is intended to demonstrate. There is no natural number standing between zero and one. Who would doubt it? Yet it must be shown, and shown step by step. Difficult ideas are required.

Why bother?

A tricky bargain is inevitably involved. In mathematics, something must be invested before anything is gained, and what is gained is never quite so palpable as what has been invested. It is a bargain that many men and women reject.

Why bother indeed?

The question is not ignominious. It merits an answer.

In the case of many parts of mathematics, answers are obvious. Geometry is the study of space, the mysterious stuff between points. To be indifferent to geometry is to be indifferent to the physical world. This is one reason that high-school students have traditionally accepted Euclid with the grudging sense that they were being forced to learn something that they needed to know.

And algebra? The repugnance (in high school) that this subject evokes has always been balanced by the sense that its symbols have a magical power to control the flux and fleen of things. Farmers and fertilizers were the staple of ancient textbooks. But energy and mass figure in those that are modern. Einstein required *only* high-school algebra in creating his theory of special relativity, but he required high-school *algebra*, and he would have been lost without it.

Mathematical analysis came to the mature attention of European mathematicians in the form of the calculus. They understood almost at once that they had been vouchsafed the first and in some respects the greatest of scientific theories. To wonder at the importance of analysis, or to scoff at its claims, is to ignore the richest and most intensely developed body of knowledge acquired by the human race.

Yes, yes. This is all very uplifting, but absolutely elementary mathematics? Not very long ago, the French mathematician Alain Connes invented the term *archaic mathematics* to describe the place where ideas are primeval and where they have not yet separated themselves into disciplines. It is an elegant phrase, an apt description. And it indicates just why elementary mathematics, when seen properly, has the grandeur of what is absolute. It is fundamental, and so, like language, an instinctive gesture of the human race.

A *theory* of absolutely elementary mathematics is an account in modern terms of something deep in the imagination; its development over the centuries represents an extraordinary exercise in self-consciousness.

This is what justifies the bothering, the sense that, by seeing an old, familiar place through the mathematician's eyes, we can gain the power to see it for the first time.

This is no little thing.

—*Paris, 2010*

1

One sheep, two sheep, three. Wool to follow . . .

NUMBERS

The natural numbers 1, 2, 3, . . . , play a twofold role in our ordinary affairs. Without them, there could be no counting, and so no answer to the question *How many?* A man who is unable to tell whether he is looking at one sheep or two of them cannot *identify* sheep. He is left staring at so much wool on the hoof. It is the natural numbers that offer him relief from sheeplessness. "The creation of numbers," Thierry of Chartres remarked in the twelfth century, "was the creation of *things.*"

As counting endows things with their identity, so it imposes on them their difference. Three sheep make for three *things.* The natural numbers are the expression in nature of division and distinctness. Between the number one and the number two there is, after all, nothing whatsoever, and nothing between things that are distinct either, however much alike they might be in various respects. The discreteness of the natural numbers is as absolute as the one enforced by the surface of our skin, which permits contact but not, alas, commingling.

There are certainly substances in the world that cannot be counted—mud, for example. The word 'mud' seems indifferently to designate mud wherever it is and however it may be found. But so strong is the intellectual impulse to subordinate experience to counting that ordinary English provides the tools by which even mud can be made numerate—a *spot* of mud, or a *dab*, or a *pile*, whereupon there is *one* spot, *two* dabs, or *three* piles. The same *one*, *two*, and *three* used in counting sheep are also used in sorting them. It is the natural numbers that make it possible for some leather-faced Spanish sheepherder, his concave cheeks pursing around two gold teeth, to put his sheep and thus his life in order.

The *first* is mine, *hombre*, as these sheepherders say, the *second* yours, and the *third* his.

A SCRIBAL ART

The Sumerians drilled their children in **AEM** more than five thousand years ago, when the desert sun was new and nothing was yet old. Sumerian children were taught the basics; their teachers had grasped the essentials. They did not find it easy. Sumerian scribes studied for years beyond childhood in order to hen-scratch clay tablets with tax records, business claims, legal codes, real-estate transactions. They left behind a sense of their mathematical intimacy, the first in history.

An inadequate sense of their calling was not among their afflictions. "The scribal art," one wrote, "is the father of masters."

Scribes *alone*, he added, could "write [inscribe] a stele; draw a field, settle accounts."

There is a gap in the text, a break in its flow.

And then a phrase isolated on both sides, one suggestive of the scribe's intellectual grandeur: ... *the palace* ...

At the end of the third millennium B.C., the Sumerian empire ran streaming into the desert sands, defeated at last by time. Carried by the wind, I suppose, or some other current of warm thought, the scribal sense of intimacy with **AEM** was acquired by Chinese mandarins, intoxicated by their new power over pictograms, and acquired again by the Babylonians, so that it appears throughout the ancient world.

Different societies used **AEM** in their own ways and for their own ends. Every society missed something, and no society, not even our own, knew or knows it all.

A MAN APART

Leopold Kronecker was born in 1823, his birthplace, the small city of Liegnitz, in East Prussia. Having quaked to the sound of Russian tanks in late 1944, Liegnitz is now known as Legnica. It is a part of Poland. East Prussia has vanished.

Kronecker's face is not easy to read on a photograph. Harsh lighting and extended exposure have darkened and deepened every facial line. The stern creases suggest an unacknowledged blood tie between Leopold Kronecker and General William Tecumseh Sherman. In both men, the forehead is high, and the hair cut short, almost *en brosse*; the eyes are deeply recessed and melancholy. In all this, Kronecker, at least, is completely Prussian and austere, but his nose has undertaken a racially unmistakable life of its own, hooking proudly at the bridge, and then curving toward its fluted tip.

I mention this not in order to make fun of another man's nose—I have a give-away nose all my own, after all—but to convey something of Kronecker's capacity to stand apart from other mathematicians while standing among them. Kronecker was that rare character in the history of thought, a *mathematical* skeptic, unwilling to countenance ideas that he could not completely grasp, and very quick to conclude that he could not grasp most ideas completely. If Kronecker the Glum was notable for saying *no*—*no* to the negative numbers, *no* to the real numbers, *no* to sets—he was notable for saying *yes* to the natural numbers, a great life-affirming *yes* to these ancient objects of thought and experience, a *yes* spilling over to encompass any mathematical construction that returned to the natural numbers in a finite series of steps.

Kronecker, the man of a thousand *no*'s, and Kronecker, the

man of a single *yes*, were combined in a singular personality: suave, supple, self-satisfied.

While still in his twenties, Leopold Kronecker pursued a career in business as the manager of his uncle's estates in East Prussia. He had a remarkable gift for practical affairs, and over the course of eight years, he made himself a wealthy man. Thereafter, he bought a splendid Berlin mansion, and after marrying his uncle's daughter, Fanny Prausnitzer, made it a center of culture and refinement.

Wealth made Kronecker indifferent to the great game of mathematical chairs in which the leading mathematicians of Europe stood looking eagerly at a small number of seat cushions still glowing with the warmth of some departed professor's buttocks. When the music stopped, they scrambled unceremoniously for the vacant seat. Inevitably, most were disappointed. Mathematicians of genius, such as Georg Cantor, spent years waiting for a call from Berlin and were bitterly vexed when it did not come.

Herr Kronecker expressed no very great interest in becoming Herr Professor. He did not need to scramble for his seat. Or for his supper. What he lacked was the right to lecture at the University of Berlin. This he wished very much to have. Devoted to topics in number theory, elliptic functions, and algebra, his papers were in every way remarkable without in any way being revolutionary. When he was elected to the Berlin Academy in 1861, he gained the right to lecture at the university.

Having declined to mount the greasy pole, he found himself at its very top. Once there, he determined to persecute those with whom he disagreed. It was an activity he carried out with never-flagging zeal.

IN EVERY HUMAN MIND

At the very beginning of human history, a Neolithic hunter chipped a number of slash marks or tallies onto his ax handle.

Was he recording bison killed? I do not know. I like to think that as *my* ancestor, he had a contemplative nature, and regarded the numbers as things in themselves, leaving those bloated bison to his rivals.

If the natural numbers appear at the very beginning of human history, they also appear spontaneously in every human mind. Otherwise, arithmetic could not be taught. Anthropologists are often amazed by the radically incommensurable way in which different societies organize the most basic facts of experience. Seeing this is said to be one of the pleasures of travel. Nonetheless, our own *one, two, three,* the Latin *unus, duo, tres,* and the Akkadian *dis, min, es,* designate precisely the same numbers. If goat eyes are a delicacy in Khartoum but not in New York, it is nonetheless true that three goat eyes is one more than two in both cities.

Because they are universal, the natural numbers very rarely are the cause of introspection. We take them for granted. We would be lost without them.

There they are.

What they are is another matter entirely.

The English logician and philosopher Bertrand Russell was a passionate opponent of the First World War, and he took the occasion of his confinement as a conscientious objector to organize his thoughts about the nature of the numbers. It may be imagined that Russell was writing under conditions of personal austerity, but in his *Autobiography* he indicates that, except for his freedom, he was provided every comfort by his jailers.

The book that Russell wrote in prison, his *Introduction to Mathematical Philosophy,* is a work of logical analysis. It has had a very great influence among mathematicians and philosophers, because it offers an account of the natural numbers in terms of something *other* than the natural numbers. Such an account is needed, Russell believed, because the numbers are "elusive" in their nature, and though they make their influence felt in the most ordinary of activities—counting sheep, after all—what they are doing is far easier to determine than how they are doing it.

For one thing, the numbers are not physical objects. They are not objects at all. Three sheep are in the pasture. There are not, in addition to the sheep, three numbers loitering around and munching grass.

Nor, however, are the numbers properties of physical objects. Three sheep are three in number, just as they are white in color. This is a step in the right direction. But to argue that being three is just like being white invites the question of just what property makes three sheep three. We know what makes them white: it is their color. To say that what makes them three is their *number* does not seem a step forward. If we knew what the numbers were, we would know what three of them add up to.

To the question of just what makes three sheep three, Russell argued that those three sheep were similar to other sets of three things—triplets, troikas, or trios. This is obviously so. Three sheep and three shepherders *are* alike. There are three of them. Russell next argued that being alike in being three could be defined with no appeal to the number three. This is the crucial step. Three sheep and three sheepherders are alike if each shepherder can be matched with one and only one sheep, and vice versa. Numbers are not required. No sheep lacks a herder and no herder a sheep.

This is ingenious, but it is also disappointing. The number three was destined to disappear in favor of similar sets, but what, after all, makes a set of three sheep a set of *three* sheep and not four? Four sheepherders and four sheep may also be put into correspondence so that nothing is left over and no sheep or sheepherder left out. The obvious answer is that a set of four sheep is larger than a set of three sheep.

It is, in fact, larger by precisely *one* sheep.

.   .   .

13

WIVES, GOATS, NUMBERS

The natural numbers begin at one; they increase by one; and they go on forever. The knowledge that this is so is an inheritance of the race. Anthropologists, it is true, report that certain tribes lack a complete sense of the numbers. Men count in the fashion of *one, two, many,* and they refer to any number past two by observing glumly that it is many. *One chief, two goats, many wives,* as a great chief might say.

I am skeptical of such reports, because I feel quite certain that stealing one of the chief's wives would prompt the chief to observe that he has *one less* than many wives. If he is capable of determining that he has *one less* wife than he might need, he is equally capable of determining that he has *one more* wife than he might want. Pressed thus by the exigencies of tribal life, he could count up by enumerating his ensuing discontent: *many wives, one more than many wives, one more than one more than many wives,* and so on toward frank domestic nightmare.

Going in the other direction, he could count down until he reached bedrock in the number one, whereupon he could compare the number of his squabbling wives with the number of his chiefs—one in both cases. It is a laborious system, to be sure, but brain workers are often indifferent to practical concerns.

GOD'S WORK

If it is difficult to say what the numbers are, it is difficult again to say how they are used. The most familiar way of counting a small number of sheep is to match those sheep to the tips of one's fingers as they uncoil from a fist; it is what we all do when we are minded to count sheep. But as an explanation of counting sheep, it suffers the drawback that counting one's fingers, however familiar, is no easier to explain than counting one's sheep. Shall we explain counting three sheep by an appeal to counting sheep one at a time, with counting by one explained in terms of a physical act—moving those sheep one by one from pasture to paddock, for example? It has an appeal, this sort of explanation. Something is being done, and when it has been done, someone has done something. But obviously, if we wish to understand what it means to count three sheep, it will hardly improve our grasp of the matter to be told that we must first count sheep one by one *three* times. What holds for counting holds for ordering too, as when a shepherd's ever-useful fingers are used to explain the fact that the first sheep into the paddock comes *before* the second, and the second *before* the third. Fingers are vigorously extended: the first, the second, and the third. Yet, if sheep are following in a certain order, it is hardly to the point to appeal to the *same* order among fingers as among flocks. If the order is not the same between fingers and flocks, of what use are the fingers? If it is the same, of what use the analogy?

At a certain point—*now*, perhaps—it becomes reasonable to suppose that neither the numbers nor the operations they make possible permit an analysis in which they disappear in favor of something more fundamental. It is the numbers that are fundamental. They may be better understood; they may be better described; but they cannot be *bettered*.

The natural numbers, Leopold Kronecker remarked, are a gift from God. Everything else is the work of man. This is a radical position in thought—an admission, on the one hand, that the natural numbers cannot be explained, and a suggestion, on the other, that the mathematician's proper work must be to accept this strange gift and from it derive all others.

It is comforting to realize that in **AEM** we are doing God's work.

Absolute Elementary mathematics

2

Henry had six wives but 'Henry' has five letters. There is a distinction between numbers and their names. Without the distinction, it is difficult to understand how numbers are named and impossible to understand the ancient and civilized art of positional notation.

POSITIONS IN POWER

The distinction is not an easy one to grasp, even for mathematicians, and it is a difficult distinction to maintain, even for me. What was it that Horace said? Even Homer nods. In 'Henry' has five letters, it is the name that counts; in Henry had six wives, the man. Logicians and philosophers use single quotation marks to specify Henry's name, but in this book, boldface is at work to the same end—**Henry** versus Henry, and **1** versus 1.

The distinction between names and numbers often slips, and with it the mathematician.* Thus in a very interesting book entitled *What Is Mathematics, Really?*, Reuben Hersh defines the relationship of equality between numbers by an appeal to the formulas in which they are named. Two *numbers* are equal, Hersh writes, "if in any *formula*, [one] may be replaced by [the other] and vice versa."†

---

* In this book, this rule: When referring *obviously* to symbols, boldface; when referring *obviously* to what they name or designate, regular face; and when there is a latent ambiguity between the two, one disambiguated by context, regular face again. Readers anticipating a completely consistent presentation, I should say at once, are apt to be disappointed. Short of a treatise, there is no way to provide one.

† Writing in the May 13, 2010, issue of *The New York Review of Books*, and assigning himself a fine eye for detail, John Paulos observed that "whenever I see the bumper

This could not be true; it is not so. Formulas are symbolic forms: marks on paper, sounds on the moving air, or even lines within a computer program. The number four cannot replace anything in any formula. Only symbols can replace symbols.

On the other hand, the number four was equal to itself long before formulas were in existence; and equal to itself, for that matter, before the earth cooled, or the solar system formed, or the universe erupted into being out of nothingness.

Numbers owe their identity to no symbolic contrivance. They are what they are. They have always been what they always were. They are not destined to change. But the numerals (and so the formulas) name; they denote; they designate; they are a part of the apparatus with which we make up a world of symbols in order to represent a world of things. How symbols designate numbers is mysterious, because we do not understand how names designate things. Speaking in *The Book of the Thousand and One Nights*, a Prince offers as good an explanation as any: There is no letter in any language, he remarks, "which is not governed by a spirit, a ray, or an emanation of the virtue of Allāh."

A MAN FOR ALL SYMBOLS

The notation used to name the natural numbers is Hindu-Arabic, and it seems to have gained currency among mathematicians during the early part of the ninth century.

The man most associated with the transmission of the Hindu-Arabic numerals to the west is Abu Ja'far Muhammad ibn Musa al-Khwarizmi. Born at some time during the latter part of the eighth century, and dying at some time during the middle of the

---

sticker 'War is never the answer,' I think that, to the contrary, war most certainly is the answer, if the question is 'What is a three-letter word for organized armed conflict?' "

But war is, of course, not a three-letter word; it is not a word at all.

ninth, al-Khwarizmi was one of those superbly accomplished scholars whose life remains much improved by an incomplete biographical record. Although he wrote in Arabic, he may have been Persian in origin; some authorities suggest a Zoroastrian affiliation. A contemporary stamp with his likeness depicts a turbaned character with a long, severe face, the nose aquiline in the spirit of great curved mathematical noses, and a curly beard descending in tight ringlets. Another stamp depicts a completely different face, round, merry, and shrewd.

It is chiefly his treatise on algebra, *The Book of Restoration and Equalization,* for which al-Khwarizmi is remembered by mathematicians, and it is his chapters on the Hindu system of positional notation by which he is remembered universally, for al-Khwarizmi was the coordinating link between Indian and Arabic mathematics, one of those irreplaceable men capable of facing in two directions at once.

It is al-Khwarizmi's system that we now use, and when he introduced it to his colleagues, and to the world, in the early part of the ninth century, he told them that it embodied, "the richest, quickest calculation method, the easiest to understand, and the easiest calculation method to learn." It is supremely useful, he added, "in cases of inheritance, legacies, partition, and trade."

All homage, then, to al-Khwarizmi, a man for all symbols and so a man for all seasons.

POSITIONAL NOTATION

The Arabic numerals comprise the nine sensuous shapes 1, 2, 3, 4, 5, 6, 7, 8, and 9. The numerals from one to nine are both basic and primitive—basic because we need something with which to start, and primitive because as symbols they cannot be decomposed into anything simpler.

What is yet lacking is a way in which these symbols may be used to denote the numbers beyond nine. Left to their own devices, of course, nine numerals can hardly do more than name nine numbers.

An especially ingenious Baghdad merchant might have designated the number ten by 9 + 1. The number after that could be designated by 9 + 1 + 1. This would hardly have been a contribution to a merchant's practical concerns, if only because a bill for one hundred and seventy drachmas would have run on for pages.

The solution to this problem emerged in stages, and it emerged in the way solutions so often emerge in mathematics, a slapped-together strategy followed by a carefully contrived cleanup. Thus merchants with business more pressing than mathematical notation long ago expressed their bills of lading or sale by writing 1 plus X for the number ten, or simply 1X, where the symbol 1, by means of its new and unfamiliar *position*, designated the number ten, and where X served simply as a placeholder, a symbol signifying that nothing was being added to ten.

Thereafter, precisely the same scheme could encompass the numbers that followed, with the number eleven expressed as 1X plus 1, or 11. Having the sense, I am hoping, that he might have discovered something profound, that Baghdad merchant might well have noted with satisfaction that by writing 1X for ten, or 11 for eleven, he had discovered the key to positional notation, the great door swinging open to admit both merchants and mathematicians.

In any compound numeral of the form **ab**, where **a** and **b** stand in for the numerals from 1 to 9, position is king and key, indicating *both* that **b** is to be added to **a**, and that **b** marks things in units of one, while **a** marks them in units of ten. That bill of lading now emerges just as that long-dead merchant might have written it:

Ψ

In the name of Allah the Most Powerful, the Most High, the Most Merciful

Dates:	17 drachmas
Oil:	13 drachmas
Almonds:	1X drachmas
Figs:	1X drachmas

whereupon positional notation and the ever-useful X come into play again in the total of 5X drachmas.

The placeholder **X** was in time replaced by the symbol **0**, so that **5X** emerges in its modern and familiar form as the numeral **50**.

The cleanup followed years later when the placeholder underwent a further promotion, so that 0 is now regarded as the name of a number in its own right. And for every good reason. The simplest of hygienic routines demands it. The sum of 2 and 0 is 2. Treating 0 as a placeholder, and so as a symbol, renders this identity incoherent. A placeholder cannot be *added* to a number, any more than a horse's name can be entered in a race.

Yet the promotion of 0 from a placeholder to the name of a living number is itself hardly a model of logical scrupulousness, for if 0 is a name, just *what* does it name? The obvious answer, in virtue of the fact that two plus zero is still two, is that it names *nothing*.

But if zero names nothing, then it is difficult to make sense of *adding* zero to two. There is no *adding* nothing to anything.

If, on the other hand, 0 names something, then it is again difficult to see why two plus something should *remain* two. It hardly helps to insist that zero is that unique something that behaves as if it were nothing. Mathematicians have traditionally resolved these difficulties by embracing the thesis that at times something is nothing, a metaphysical achievement that cannot be said to inspire a sense of serenity.

The converse, that at times nothing is something, is, of course, among the most useful declarations of the human race.

In Indian thought there is a connection between zero and *shûnya*, a term meaning the emptiness, the nonexistent, the nonformed, and the noncreated. Curiously enough, zero also seems to have been associated with the infinite, the god Vishnu's foot, and a voyage on water.

In the early years of the nineteenth century, a number of English mathematicians still regarded zero with unease, and the negative numbers with distaste. Had I been among them, I would have embraced their cause. It may not be too late.

3

The urge to get to the bottom of things, if things have a bottom, is not unique to physicists. Why accept the numbers as fundamental if there might be something more fundamental still? Why indeed?

SETS

The introduction of set theory at the end of the nineteenth century persuaded many mathematicians, Bertrand Russell among them, that they had discovered a system by which the natural numbers could be displaced in favor of something more fundamental. The creation of Georg Cantor, set theory is the most remarkable single achievement of nineteenth-century mathematics, so much so that David Hilbert was moved to call it a paradise. Hilbert was a great mathematician and a careful German prose stylist, his word of choice conveying a very precisely blended mixture of admiration and regret.

Sets are part of a *volkisch* family: troops, tribes, groups, ensembles, even prides, flocks, rabbles, ribbles, gaggles, and gabbles. These words all are in the end synonyms for *set*, or they depend on the concept of a set for their coherence, a ribble of rabbis, a set of them or a collection, but in any case something more than a rabbinical heap. Beyond saying that a set is any real or potential object of thought, Cantor very wisely said nothing more.

Sets by their nature have members, things that belong to them. Membership is the fundamental relationship of the theory. No relationship could be more primitive than one answering to the interrogative whether something is in or out.

There is in the idea of a set plainly an extraordinary degree of freedom. This freedom endows even the simplest of sets with a dangerous reproductive urgency. Having for so long lived lives as sheep, three sheep must now be imagined as a set-theoretical flock, a fact that mathematicians symbolize by collecting their names in the pen between curled parentheses, {**Sheep 1, Sheep 2, Sheep 3**}.

If there were three sheep to begin with, and so three things, there are now *four*: the three sheep *and* the set collecting them. That set is again an object of thought.

If this is not so, just *what* have we been thinking about?

The universe now includes the set {{**Sheep 1, Sheep 2, Sheep 3**}}, whose sole member is the set of three sheep whose members in turn are those damned sheep. There are *five* objects in the universe where just a moment ago there were only a handful of sheep.

This is a process that may be indefinitely continued. Sets are profligate; they multiply by iteration. Nor is a uniquely mathematical process at work. It is hardly mathematical at all. Commenting on a religious audience in the city of Qom, V. S. Naipaul remarked that "faith like this . . . ," and then because he lacked a word to describe how faithful the faithful were, he added, "faith in the faith," and so suggested that faith, like the operation by which sets are formed, may be self-applied, faith in the faith being distinct from faith itself.

Something of madness attaches to these iterations, because they suggest no standard of control. Is there for the faithful faith in the faith in the faith?

"Successful indeed are the believers," as the Qur'an observes enigmatically in Sura 23.1.

ONE OF A KIND

The set of sheep has sheep as members; the set of sumo wrestlers, sumo wrestlers; but as for sheep who are also sumo wrestlers, there are none. The set is empty, thank God. But while it is empty, *it* is not nothing.

Quite the contrary. Sets are abstract objects, capable of surviving a radical decline in membership. A heap of sheep may be diminished one sheep at a time until both the sheep and the heap are gone. A heap is nothing more than its sheep. A *set* of sheep, however, survives sheeplessness to emerge on the other side as a set with no members.

Sets with no members are sisters, as far as set theory goes: nothing belongs to any of them. If two sets are identical because they have the same members, then the empty sets are all the same because none of them has any. Membership hardly comes dearer than this. No club is more exclusive. It follows that there is only one empty set, and that is *the* empty set, which mathematicians designate as $\emptyset$, the symbol looking very much like a blindly staring eye, its powers canceled.

Questions about zero now resolve themselves in favor of the artifice that the empty set corresponds to zero. "It's like a party with no guests," a student once remarked.

It is just like that. Merchants are content, and mathematicians too, for without zero there is no moving merchandise and certainly no doing mathematics.

If zero corresponds to the empty set, the number one can find purchase as the set containing just the empty set, or $\{\emptyset\}$. It has one element, after all. The number two is in the same way identified with the set containing the empty set and the set containing the empty set, making two: $2 = \{\emptyset, \{\emptyset\}\}$. Thereafter all of the natural numbers may be constructed as a tower of *sets* instead of a tower of numbers, so that every natural number corresponds to a particular set.

This shows that the numbers may be paired to the sets but hardly that the sets are more fundamental than the numbers. To see that the set $\{\emptyset, \{\emptyset\}\}$ comprises two members, it is necessary first to count them. Counting sets is an undertaking as dependent on the natural numbers as counting sheep.

PARADOX

Within Cantor's lifetime, logicians demonstrated that set theory is frankly inconsistent. Dangers in mathematics do not get more dangerous than this.

Of the paradoxes, Russell's is the most famous; and it is also the easiest to state. Some sets are members of themselves, and some are not. The set of all sets is again a set: it is an object of thought. But the *set* of all sheep is not a sheep even though it is a set.

What, Russell asked, of the set of all sets that are *not* members of themselves?

Is *it* a member of itself?

It is if it isn't, and it isn't if it is.

This is not a conclusion that anyone might find encouraging, least of all mathematicians otherwise well satisfied that theirs is a discipline in which consistency is championed.

In 1908, Ernst Zermelo proposed a set of axioms for set theory, and suggested hopefully that, as long as they were respected, all would be well.

To this day, no one really knows whether this is so. A *proof* is unavailable.

And all this for zero, which is to say, for nothing.

4

*Human knowledge is radically unstable. We are strang-
ers to one another and, more often than not, to ourselves.
In telling you that you do not know what you think you
know, I am not telling you anything that you do not know.*

CERTAINTY

And yet mathematics offers the impression of certainty, a shrug of
the absolute. No one raises a skeptical finger in the air on learning
that if five is greater than four plus zero, then it is also greater than
four. It is an impression at odds with human life, and radically at
odds with the other sciences.

Let me offer an example. In the second century A.D., the
Greek mathematician and astronomer Claude Ptolemy created a
comprehensive astronomical theory. The heavens he imagined as
a great sphere, the earth at its center. Ptolemy's masterpiece is
entitled the *Almagest*, the word meaning "the greatest" in Arabic
and in the Greek from which the title is taken. The *Almagest* is an
attempt to grasp the universe in mathematical terms. It is the first
such attempt in history, and for this reason, the greatest, just as its
name suggests. Although Ptolemaic astronomy is often thought
false, Ptolemy and Johannes Kepler are secure in their position as
the greatest among the greats. There is no third.

It is in the second part of the first book of the *Almagest*, entitled "On the Order of the Theorems," that Ptolemy describes his
ambitions. They are considerable; they are grandiose. "In the treatise which we propose," he writes, "the first order of business is to
grasp the relationship of the earth taken as a whole to the heavens
taken as a whole." The second order of business is to describe

"the motions of the sun and the moon," and the third, to give an account of the stars. Ptolemy then summarizes his conclusions. "The heavens [are] spherical in shape and move as a sphere; the earth, too, is sensibly spherical in shape, when taken as a whole; in position it lies in the middle of the heavens, very much like its center; in its size and distance, it has the ratio of a point to the sphere of the fixed stars, and it has no motion from place to place."

"Absolutely all phenomena," Ptolemy adds somberly, "are in contradiction to any of the alternative notions that have been proposed."

NOTHING HUMAN IS CERTAIN

For fifteen hundred years, the *Almagest* appeared as solid and as enduring as pig iron, the theory that it expressed brilliantly meeting the demands of new astronomical data, such as the retrograde motion of the planets, by means of an elaborate system of epicycles and deferents. Until well into the seventeenth century, the advantages of the Copernican system were not clear, and its disadvantages considerable. Copernican astronomers could not plausibly explain why, if the earth were in motion around the sun, no one on its surface ever noticed.

And yet Ptolemaic astronomy was discarded soon afterward, and thereafter ridiculed for the very techniques that lent it accuracy. It is today an object lesson, and so a warning, the facts in the end turning against the theory:

> —The earth is not at the center of the solar system.
> —The sun does not move.
> —The planets do not describe circles in the sky.
> —The heavens are not spherical.

Nihil homini certum est, as Ovid observed. Nothing human is certain.

AN EXCEPTION TO OVID

Mathematics is the great exception to this melancholy observation; and Ptolemy is, in this regard, a one-man multitude. A number of powerful theorems in geometry rise from the ruins of his system, like tendrils forcing themselves through the rubble and toward the sun.

A circle is given in the plane and within the circle a four-sided figure inscribed. The sum of the products of their sides, Ptolemy demonstrated, is equal to the product of the two diagonals. This result is today known as Ptolemy's theorem.

The contrast between the man of astronomy and the man of mathematics is poignant. Ptolemy had invested his hope for glory in his theory of the heavens; the mathematics that he developed, he regarded as an instrument.

The glory was there all along, but it was not where Ptolemy had thought to find it.

What accounts for the difference between the man of mathematics and the man of science?

The popular view is that in mathematics proof is possible, and beyond mathematics it is not. "As far as the laws of mathematics refer to reality," Einstein remarked, "they are not certain, and as far as they are certain, they do not refer to reality." Of this view, the most that can be said is that it is bizarre. We take reality with our eyes wide open. Why should what we take in with our senses be *less* certain than what we conjure up in saying that, if four is greater than three and three greater than two, then four is greater than two? It would seem to be quite the other way around.

No doubt there are proofs in mathematics. Proofs are the mathematician's coin in trade. The question is why there are none *elsewhere*.

A proof, after all, is a mathematical *argument*, and so a part of an old and familiar human genre. The logician, and not the mathematician, is in charge. It is the opportunity of a lifetime. Beyond mathematics, the logician has, in any case, a very substantial portfolio. Since his business is conveying premises correctly to their conclusions, his subject is relevant to any activity that arrays human beings against one another or against themselves: domestic disputes; financial controversies; conflicts about abortion, family life, corporate organization, international law, simple decency, flag burning, homeopathic medicine, ancient archaeology, women's rights, the rules of warfare, dress codes, intelligent design, conspiracy theories, Freudian psychology, or anything else that may be plausibly embedded in a continuum ranging from *Honey, Let's Not Fight Anymore* to *Exterminate All the Brutes.*

And yet, across the vast range of arguments offered, assessed, embraced, deferred, delayed, or defeated, it is only within mathematics that arguments achieve the power to compel allegiance because they are seen to command assent.

No philosophical theory has ever shown why this should be so. It is a part of the mystery of mathematics.

THE GREATEST

Aristotle was the first and the greatest of logicians, with even Kurt Gödel judged great (by J. Robert Oppenheimer, among others) because he was the greatest logician *since Aristotle.* Entitled the *Organon,* Aristotle's work on logic is virtually the only part of his remarkable corpus to be reliably traced back to his own hand.

The eighteenth-century English philosopher Thomas Reid has offered a shrewd account of Aristotle's genius. Aristotle, Reid notes, "had very uncommon advantages." He was born in Greece "in an age when the philosophical spirit in Greece had long flourished," and was for "twenty years a favourite scholar of Plato, and tutor to Alexander the Great, who both honoured him with his friendship, and supplied him with every thing necessary for the prosecution of his inquiries." These advantages, Reid goes on to say, "he improved by indefatigable study, and immense reading." And as to his genius, "it would be disrespectful to mankind not to allow an uncommon share to a man who governed the opinions of the most enlightened part of the species near two thousand years." This is both handsome as a tribute and correct as a compliment. Determined to find some criticism to offer Aristotle, the sober Reid can think to say only that he was human, for he "seems to have had," Reid argues, "a greater passion for fame than for truth, and to have wanted rather to be admired as the prince of philosophers than to be useful."

WHAT IF, WHAT THEN

To Aristotle is due the foundational insights of logic itself, the keys to its nature. There are just two. An argument is valid in virtue of its form and not its content; and the validity of an argument is conditional and so a matter of asking *what if* and seeing thereafter *what then*.

Here is an argument offered by the American logician Alonzo Church (in the introduction to his treatise, *Introduction to Mathematical Logic*):

The first premise:	Brothers have the same surname.
The second:	Richard and Stanley are brothers.
The third:	Stanley has the surname Thompson.
The conclusion:	Richard has the surname Thompson.

And here, Church adds, is quite another:

The first premise:	Complex numbers with real positive ratios have the same amplitude.
The second:	$i - \sqrt{3}/3$ and ω are complex numbers with real positive ratios.

The third:	ω has amplitude $2\pi/3$.
The conclusion:	$i - \sqrt{3}/3$ has amplitude $2\pi/3$.

In both arguments, premises are passing toward their conclusions, but dealing as it does with a branch of mathematics known as complex analysis, the second argument might well have been written in Warlpiri; it has nothing to do with the Thompsons—heirs to a submachine-gun franchise, as I recall—or with **AEM**.

Nonetheless, these two arguments have the same form, and the fact that the second argument is unintelligible does not in the least detract from its validity.

Whereupon the second Aristotelian insight: validity is conditional, a fact that we recognize by appealing to what is not the case *for the sake of argument.* A valid argument is such that *if* its premises *were* true its conclusion *would* be true as well, a counterfactual *(were it so)* and a modal imperative *(it would be so)* now commanding the discussion. A valid argument does nothing to guarantee the living truth of its premises. Logic is the logician's business; truth he leaves to others.

Here, again, is Church's first argument:

1

The first premise:	Brothers have the same surname.
The second:	Richard and Stanley are brothers.
The third:	Stanley has the surname Thompson.
The conclusion:	Richard has the surname Thompson.

And here is the same argument without the pad of its premises, a cannonade of *ifs* converging to a brotherly *then:*

2

If all brothers have the same surname, and *if* Richard and Stanley are brothers, and *if* Stanley has the surname Thompson, *then* Richard has the surname Thompson.

If it is perfectly plain that 2 captures the movement of inference in 1, and vice versa, there is nonetheless within mathematical logic a proof that this must be so. Known as the deduction theorem, it says what one might expect: that 1 and 2 come to the same thing.

The deduction theorem ratifies what is obvious; but the ratification should not be allowed to prompt a secondary confusion.

Yes, the deduction theorem affirms, to trust in 2 given 1; but equally No to trust in its conclusion standing alone.

Who knows whether Richard and Stanley *really* are brothers? And if no one knows, then who knows whether Richard and Stanley *really* have the same surname?

Not the logician.

THE FULCRUM

There is no more chilling account in the histories that Herodotus offers than his story of the Lydian King Croesus. Alarmed by Persian power, as sinister then as it would seem to be now, he contemplated a pre-emptive attack. Before he gathered his forces and secured his alliances, he consulted the Oracle at Delphi and asked whether he would prevail in battle.

He received this response:

> *If* you attack the Persians, *then* you will destroy a great empire.

Made optimistic by the Oracle's assurances, Croesus attacked the Persians. He was defeated, and surrendered in slavery. The great empire he destroyed was his own.

Hypothetical statements play a crucial role throughout **AEM**. And not only throughout **AEM**. They form an indispensable inferential bridge throughout the law, literature, and life:

If A and B agree that A will act as B's secretary for one year at a salary of $100 a week, **then** *the contract is said to be divisible.* [*Calamari and Perillo,* Contracts.]

If you couldn't be happy with her, **then** *why on earth should you expect to be happy with anyone else?* [*Evelyn Waugh,* Brideshead Revisited.]

If, by chance, a photon successfully traverses such a filter, **then** *the chance it will get through a second filter with exactly the same orientation will be 100 percent.* [*John Maddox,* What Remains to Be Discovered.]

Some propositions make their case in one fell swoop—*Your check is in the mail* saying just what needs to be said to one's grasping creditors, and no more. A single proposition is involved, and it is either true or false. Hypothetical propositions, on the other hand, make their case in two swoops instead of one, for their truth depends on the truth of their constituents, the hinge of fate creaking twice. If you attack the Persians—that is one proposition. Then you will destroy an empire—and that is another proposition.

With two propositions in play there are four possible ways in which they could be jointly true or false. Both could be true; both false; or the antecedent could be true and the consequence false; or the reverse. Letting *P* and *Q* stand for arbitrary propositions, there is

IF P THEN Q

P is true	T	*Q* is true
P is false	T	*Q* is false
P is true	**F**	*Q* is false
P is false	T	*Q* is true

It is the third possibility that logicians flag as compromising the truth of the hypothetical as a whole; in every other case, the hypothetical proposition is counted true. Sensible enough. Hypothetical propositions are a conveyance. They go *from* their antecedent *to* their consequent. If starting out in the right way—their antecedent is true—they nonetheless fail to get to the right place—their consequent is false—they do not get where they are going, and must therefore be dismissed. This means, of course, that hypothetical statements are true if their antecedent is false. This very often prompts a susurrus of discontent. Could a hypothetical proposition be true *just* because its antecedent is false? It hardly seems fair. Logicians have never argued the point, perhaps because the only argument they *could* offer is disappointing in its modesty. Would intuition be better served, they might say, were hypothetical propositions reckoned *false* if their antecedents were false? It seems hardly an improvement.

The particles *if* and *then* are in modern logic known as propositional (or sentential) connectives—*connectives* because of what they do, and *propositional* because of whom they connect. There are many such connectives in ordinary language beyond the hypothetical. For purposes of logical analysis, the most important are *not* (negation), *and* (conjunction), and *or* (disjunction). Such connectives as *although, but, just in case,* and *if and only if* are variants expressing the natural exuberance of ordinary language.

The logic by which propositions are governed is not enough to control the flow of inference within mathematics, or even within **AEM.** In order to say simply that there is no natural number between zero and one, resources are needed that express the nuances of quantification. Those resources having been provided by twentieth-century logicians, the abiding mysteries of form remain.* If there is no natural number between zero and one, neither is there an elected official between the President and the

---

* See Chapter 8.

Vice President: inferences about the numbers and inferences about politicians very often turn on similar propositions.

But in mathematics, they lead to certainty; and in politics, they do not. And there is no continuum between cases either. Mathematics is a world apart, in its language, its objects, and the certainty that its proofs afford.

5

Mathematics is a cold master and logic still colder. This is the popular view; it is not far wrong.

A COLD MASTER

Pierre Abélard was born in what is now Brittany in 1079; and as the most important logician of the High Middle Ages, he falls evenly between the two great eras in the history of logic, the first taking place in ancient Greece, and the second in nineteenth- and twentieth-century Europe. His family comprised the minor nobility, and as the eldest son, he was expected to become a soldier, a career that he rejected, he writes, because he preferred "the conflicts of disputations to the trophies of war." Thereafter, Abélard was introduced to eleventh-century philosophy, chiefly by Jean Roscelin. With his education complete, Abélard wandered the Loire Valley, "disputing," as he says, "like a true peripatetic philosopher, whenever I heard there was keen interest in the art of dialectic.

"At last I came to Paris," Abélard writes. Then as now, the city radiated waves of glamour and prestige, attracting troubadours and poets, logicians and philosophers, architects, artisans, stonemasons, goldsmiths, windy prelates, moneymen eager for cathedral contracts, and a remarkable number of prostitutes, drifters, lowlifes, spongers, wastrels, petty criminals, jugglers, necromancers, astrologers, minor clergy, dissipated aristocrats, heretics, and, of course, hunchbacks.

Having drawn the circle of his own wanderings to their center, Abélard wasted no time in denouncing the views of his rural master, Jean Roscelin. Not very much is known specifically of the doctrines that Roscelin had preached. A nominalist in name, Roscelin believed in words, and so became a minimalist in philosophy. Where Plato, and so many others, saw in terms such as *red, good, brave, loyal,* and *hirsute* the names of universals or Platonic forms, Roscelin stopped at the water's edge, seeing nothing in words beyond words. Condemned for heresy in 1092, he was exiled to England, the Catholic Church having correctly noticed that once a man is disposed to doubt the existence of universals, his doubts about the Trinity could not be far behind. Abélard's own criticisms followed the longboat or skiff that carried Roscelin across the choppy waters of the English Channel, the man's sense of indignation mounting as he faced the English coast, with Abélard's criticisms stinging at his backside.

"If you had savored only a little bit of the sweetness of the Christian religion," Roscelin would later write, and there followed the usual complaints of a teacher making the pained discovery that teachers always make—namely, that their students are no longer sufficiently mindful of the "great benefits" that they have received from their instruction.

No universities—not yet. No degrees. No committees. No chairs. No tenure. Teachers themselves established schools. They clambered onto hillsides and, with their students arrayed before them, talked into the wind. Abélard considered his contemporaries perfect fools. "I began," he writes, "to think of myself as the only philosopher in the world, with nothing to fear from anyone." His reputation owed much to a contrived encounter with an older, more established philosopher, William of Champeaux, "the supreme master of the subject," as Abélard observes. William was the archdeacon of Paris and head of the Cloister School at Notre-Dame, a big, bruising, well-known figure, much crippled by his unenviable ability to express a philosophical position adroitly without in the least being able to defend it intelligibly. If Roscelin had dismissed universals from the world, William called them back, insisting with a slow, measured shake of his tonsured head that *justice, humanity, goodness, whiteness,* and *beauty* are as real as Socrates or Aristotle, the proposition that *Socrates is a man* denoting Socrates *and* designating his humanity.

A nervous coughlike harrumph erupts from the back of the lecture room. Abélard shambles to his feet. *If* Socrates is a man *and* Aristotle is a man, he asks, *then* is the *same* humanity *in* both men?

Having no real idea what to say, William says finally that, "in the common existence of universals, the whole species [is] essentially the same *in* each of its individuals. . . ." And thereupon Abélard pushes poor baffled William, his wattles now wobbling indignantly, from one absurdity to another, concluding finally that on William's view it follows inexorably that Socrates is identical to a donkey. "Although he [William] welcomed me at first," Abélard writes, "he soon took a violent dislike to me because I set out to refute some of his arguments and frequently reasoned against him."

As his contemporaries observed, Abélard was everywhere in Paris in the first two decades of the twelfth century, talking, writing, lecturing, and in general jabbing his tense index finger into a great many withdrawing and affronted chests, his logical skills now so sharp that by means of endlessly divided distinctions he seemed able to slice up the very air he breathed. It would seem that, on hearing Abélard's lecture, Anselm of Laon became "wildly jealous," a circumstance that Abélard assigned to every conceivable cause except the one that he had set in motion.

RUIN

"Since the beginning of the human race," Abélard observed with some asperity in his autobiography, *Historia Calamitatum (A History of My Misfortunes),* "women [have] brought the noblest men to ruin."

"There was in Paris at the time," Abélard writes, "a young girl named Héloïse."

Born in 1100, Héloïse grew up somewhere outside Paris, and was educated at the abbey of Notre-Dame at Argenteuil, the child she must have been disappearing into the soft folds of womanly flesh she became, the man Abélard would be emerging from the irritable adolescent he was, double pupation, I suppose, and so something that conforms entirely to Aristotle's idea of a chance event when those liberated butterflies meet in mid-flight. I could have seen the two of them from my window, Héloïse passing in quick, tripping steps in one direction, even as brown and baggy Abélard, his cassock flapping, lumbers toward her from the other, *she* tripping on, *he* executing the proverbial full stop and double take of a man whose senses are violently suffused. He was at once "all on fire with desire for this girl." Héloïse lived with her uncle, the Canon Fulbert, in a house on the quay. The original has been largely destroyed, save for a few sturdy medieval timbers, but a sign commemorates the great romantic drama that took place there long ago. The seduction proceeded by means of a series of steps meant to promote Abélard from logician in house to lover in hand. "I had," he writes, "youth and exceptional good looks as well as my great reputation to commend me." Mutual friends were invoked. They adverted to Abélard's great reputation as a teacher and his uncompromising continence. "We were united," Abélard writes, "first under one roof, and then in heart, and so, with our lessons as a pretext, we abandoned ourselves entirely to love."

Uncle Fulbert, although having never been commended for his intelligence, must at some point have noticed the moans and distracted mutters coming from *le grenier* above, with its straw pallet, sooty walls, and those small windows overlooking the brown river below. He was apparently rather slow on the uptake. Quoting Saint Jerome with satisfaction, Abélard writes that "we are always the last to learn of evil in our own home." But if Fulbert could not smell smoke, he was eventually persuaded to see fire, largely because, as Abélard remarks none too delicately but with a certain stubborn pride, he and Héloïse were "caught in the act."

"We shall both be destroyed," Héloïse remarked quite lucidly, in words that Abélard quotes. "All that is left for us is suffering as great as our love has been."

She was quite right. The passion that had consumed them then consumed them. Abélard and Héloïse were separated and then reunited. Their stories and evasions grew elaborate; and through it all, Uncle Fulbert, baffled by events and powerless to control them, fumed and steamed and plotted and schemed, until suffocating with fury, he set in motion by means of a gang of ruffians the events that would lead to Abélard's dreadful mutilation. Thereafter, both Abélard and Héloïse entered religious life, Abélard because he was unable to conceive another plan, and Héloïse because she was forced to do so by Abélard. She took her monastic vows most unwillingly. She quite knew they would bind her for life, and they did.

Abélard retired into the great coldness that had always been a part of his personality. He lived out his days by means of a series of feuds. Héloïse refused to surrender the love that had afforded her radiance: "O think of me," she wrote to Abélard, adding the plea that every lover knows, "do not forget me."

And then: "For ever adieu!"

6

The idea of an axiomatic system stands to mathematics as a whole as the idea of the Gothic cathedral stands to medieval architecture. It is the form to which mathematicians have always aspired.

AXIOMS FOR NUMBERS

In the third century B.C., the Greek geometer Euclid organized the principles of plane geometry into an axiomatic system. He was the first mathematician to think in these terms. It was his genius to do so. The axioms of an axiomatic system are its assumptions. Eggs are in one basket. From the axioms, theorems follow by the laws of logic, a proof in mathematics an austere parody of the way in which in life a real chicken is derived from a real egg. In going from axioms to theorems, the mathematician proceeds by inferential steps. What the chicken is doing, God alone knows.

Euclid accepted five axioms as the basis for his system; he urged that they be accepted without evidence. This is obviously good advice if by evidence one means the derivation of the axioms from still further axioms.

If the axioms of an axiomatic system cannot be derived from still further axioms, Euclid carelessly remarked, it follows that they must be accepted without evidence. This is not so, and it was not so for Euclid. *His* axioms, Euclid argued, were *self*-evident. Self-evidence is evidence all its own, as Euclid realized when he discovered that one of his axioms was anything but evident, even to himself.

For more than two thousand years, geometry has meant Euclidean geometry, and Euclidean geometry, Euclid's *Elements*. It is the oldest complete text in the western mathematical tradition. In every generation, a few students have found themselves ravished by its study. "At the age of eleven," Bertrand Russell recalls in his *Autobiography*, "I began Euclid, with my brother as my tutor. This was one of the great events of my life, as dazzling as first love. I had not imagined that there was anything so delicious in the world."

Until recently, a course in Euclidean geometry was a part of the universal curriculum of mankind. Training in its techniques was widely thought to improve a man's mental hygiene. With other young lawyers snoring in a common hotel room, Abraham Lincoln stayed up late in order to master Euclid's demonstrations by candlelight. The law had made his intelligence supple; Euclid made its edge hard. Students have always reported themselves better for their Euclidean experience, affirming later in life that a close study of the *Elements* imparted many virtues, chief among them the ability to think.

Nothing so conveys the way in which Euclid concentrates the mind as Euclid. The twenty-seventh proposition of the *Elements* discusses straight and parallel lines. Suppose that "the straight line *EF* falling on two straight lines *AB, CD*, make the alternate angles *AEF, EFD* equal."

"I say," Euclid says, "that *AB* is parallel to *CD*."

The proof that follows is as inviting as an immersion in cold water:

—*Suppose the proposition false.*
—*Then the alternate angles are equal to one another, but the straight lines are not parallel.*
—*Which is impossible.*

It is the reader who must grasp that by *impossible* Euclid means that the proposition he is proposing to reject leads to a contradiction; and it is the reader who must somehow see in his mind's eye the picture that Euclid is prepared to embrace; and it is the reader who must coordinate in a logical way Euclid's twenty-seventh proposition with the twenty-six propositions that Euclid has already demonstrated.

It is the reader who must do this because, shuffling off to attend to his twenty-eighth proposition, the Master has nothing more to say or add.

THE SECOND INTERNATIONAL CONGRESS

In 1900, European mathematicians held their Second International Congress in Paris. The first had been held in Zurich some years before. The mathematicians were meeting in the most beautiful city in Europe, but they were meeting in August, and as happens every summer, Parisians affirmed that they were surprised by the heat. Unfavorable comparisons were made with Swiss efficiency that Congress organizers remembered but could not duplicate.

In his *Autobiography*, Bertrand Russell describes the Congress as "the turning point of my intellectual life because there I met [Giuseppe] Peano." Born in 1858 in Italy's Cuneo Province, Peano was of peasant stock, and the only mathematician at the congress who was not a member of the middle or upper European classes. Like Enrico Fermi, he had made his way through the Italian educational establishment by means of his talent. It could not have been easy. What attracted Russell's admiration in Peano's personality was an interesting combination of traits. "In discussions at the Congress," Russell writes, "I observed that he was always more precise than anyone else." And then Russell adds a remark that both amplifies his sentiments and compromises their nobility. Peano, he recounts, "invariably got the better of any argument on which he embarked."

Giuseppe Peano made decisive contributions to the theory of ordinary differential equations; he was a well-known and influential academic; and he was something of a passionate eccentric, committed to an international scientific language of his own devising that he called *Latino sine flexione*, a kind of pidgin Latin in which case endings and inflections had all been dropped. The scheme embodied none of the merits of the Latin language and all of its defects. The late nineteenth century was an age of enthusiasm, and any number of scientists thought that if only they could persuade the scientific community to adopt a universally accessible language, all would thereafter be well. It was in this environment that Esperanto was created. I do not think that any scientist of note ever bothered to learn Peano's *Latino sine flexione*, and those who learned it never used it. Esperanto remains what it always was, and that is a language no one would think to use if not compelled to do so.

THE PEANO AXIOMS

If Euclid was the first mathematician to think in axiomatic terms, until the nineteenth century he was also the last. Over the course of more than two thousand years, no mathematician had thought to bring the numbers under the control of an axiomatic system.

In 1889, Peano published a set of axioms for the natural numbers in a little book (no more than a pamphlet) entitled *Arithmetices Principia, Nova Methodo Exposita (A New Exposition of Arithmetical Principles)*. Just why he chose to publish work of fundamental importance in classical Latin, its case endings and declensions intact, while discarding his own *Latino sine flexione*, I do not know. The ideas that Peano advanced were remarkable, but they were not original, very similar ideas having occurred to the German mathematician Richard Dedekind at roughly the same time.

There are five Peano axioms, just as there are five axioms governing Euclidean geometry:

1   0 is a natural number.

2   The successor of any natural number is again a natural number.

3   The number 0 is not the successor of any natural number.

4   If two numbers have the same successors, then those numbers are equal.

5   Any set of numbers that contains the number 0, and that contains the successor of any number *if* it contains the number itself, contains all the natural numbers.

Of these axioms, the first, the second, and the third are not the stuff from which controversy is made. They have a meaning that, if it is not evident, is at least clear.

The first axiom is intended to guard against intellectual frivolity. Perhaps there are *no* natural numbers at all? This the axiom denies. The natural numbers do not constitute a vacancy. There is at least one. No one, it is true, ever supposed the contrary, and even *imagining* a world without numbers, and so without distinctions, is remarkably difficult. What would *it* be like, the *it* itself an indication that something, and so *one* thing, is being contrasted with something else?

Sicher ist sicher, as mathematicians say: better safe than sorry.

Peano's second axiom introduces in the expression *the successor of any natural number* an undefined term. Although undefined, it has an obvious meaning. The number three is the successor of the number two because three comes after two. It comes *right after.* The expression *right after* is no more transparent than the expression *the successor of,* but in the impression of brusqueness that it conveys—*Hold on, you'll get yours **right after** he gets his*—it is perhaps more familiar.

The third axiom establishes that the natural numbers have a beginning. There is a number that does not come right after another number, and it does not come right after anything else either. The question *A beginning in what?*, because it cannot be answered, suggests that the natural numbers are like the Big Bang in marking the appearance of a complicated structure with no obvious antecedents.

Peano's fourth axiom controls the idea of succession, and with it the identity of the natural numbers. Suppose the axiom canceled or otherwise neglected. In that case, one number might be its own successor. This Peano's fourth axiom rules out.

Peano's fifth axiom stands apart from the others. On the one hand, it makes a claim: a set of numbers that meets two conditions contains *all* of them. Those two conditions? The number zero is a member of the set; and if any given number is in the set, so, too, its successor.

On the other hand, Peano's fifth axiom seems curiously akin to a legal document, one that sanctions conduct or that creates a right. It *specifies* the *conditions* under which *you* can *conclude* something about all of the natural numbers. If those conditions are not met, then *you cannot* do any *concluding* at all and had best lapse into silence.

It is Peano's fifth axiom that insinuates a troubling, suggestive, but very often hidden sequence of concepts into mathematics, one moving in an unexpected way from the natural numbers to rules about their inference. It is as if the mathematicians were with foolscap in hand imitating the lawyers, men busy studying cases and drawing general principles from them.

IL GATTOPARDO

Giuseppe Peano died on April 20, 1932, and as his American biographer, Hubert Kennedy, remarks, "He lived too long." These are terrible words, because they represent a reproach that, although widely made of others, is almost never addressed to oneself. Peano had made his great contributions to logic and arithmetic before the nineteenth century ended. He had been rewarded. He had met the leading mathematicians of his time. He had profoundly impressed Bertrand Russell.

Thereafter, something like a subtle dissolution of focus took hold of the man, one whose objective correlative, I suspect, was the increasing hoarseness that afflicted his voice, so that, if he needed to strain to make himself heard, others needed to strain to hear what he had said. At some time in the early 1890s, he had conceived of a great mathematical project. "It would be very useful," he wrote, "to collect all the known propositions referring to certain parts of mathematics, and to publish these collections." When it came to arithmetic, he proposed to publish these propositions in the logical notation that he had himself devised. His goal, it would seem, was to reduce mathematics to a very considerable list, one in which each item was logically connected to the one that came before. A sense for the meaning of the list—or *formulario*—would in theory be accessible to anyone who understood the logical notation.

The *formulario* was an exercise in self-deception, exciting the interest of no one beyond Peano's most immediate disciples, men who for one reason or another were persuaded that enthusiasm was in their best interests. Before 1900, the *formulario* was Peano's curiosity; and afterward, it became his passion. The final edition of the *formulario* Peano published in his system of *Latino sine flexione*, thus at one stroke embedding his ideas in two inaccessible symbolisms. Much to the consternation of other members of the faculty at the University of Turin, he insisted on presenting his own courses in the *formulario* style, his students quite properly complaining that they could not understand a word of what the hoarse, excitable old man was saying.

And thereafter, his life became a matter of waiting. The past came to reclaim him. He became increasingly fond of returning to his family farm in the Piemontese countryside. He dressed simply. At mealtimes, he ate what he had eaten as a child. He did not lose the things that he had known as a sophisticated European mathematician, but he came to value them less. The seasons passed.

At the conclusion of *Il Gattopardo (The Leopard)*, Lampedusa's elegy for his imagined ancestor, Don Fabrizio, the Prince of Salina, death at last arrives to claim the Prince in a stuffy hotel room. As an organ-grinder spins out melodies in the street below, the Prince, Lampedusa writes, was "making up a general balance sheet of his whole life, trying to sort out of the immense ash-heap of liabilities the golden flecks of happy moments." What remained golden was his affectionate regard for his nephew Tancredi, the memory of his dogs, his ancestral home, Donnafugata. "And why not?" he asks. "The public thrill of being given a medal at the Sorbonne." *In the growing dark he tried to count how much time he had really lived. His brain could not cope with the simple calculation anymore: three months, three weeks, a total of six months, six by eight, eighty-four, forty-eight thousand, the square root of eight hundred and forty thousand . . . And then nothing.*

At his death, Don Fabrizio, the Leopard, was seventy-three; and so was Giuseppe Peano.

7

*The Peano axioms are monumental in their accomplish-
ment because they bring the natural numbers under axiomatic control; and they are radical in their implications
because of the importance they assign to the single idea
of succession.*

SUCCESSION

In 1888, the German mathematician Richard Dedekind published a short monograph entitled *Was sind und was sollen die
Zahlen?* The title has been traditionally translated as *The Nature
and Meaning of Numbers,* but there is in the German a normative
aspect—what *should* the numbers be, and how should *we* think of
them?—that reflects a subtlety missing from the English.

For his part, Dedekind was quite clear what *he* thought about
the numbers:

I regard the whole of arithmetic as a necessary, or at least
natural, consequence of the simplest arithmetic act, that
of counting, and counting itself as nothing else than the
successive creation of the infinite series of positive integers
in which each individual is defined by the one immedi-
ately preceding; the simplest act is the passing from an
already-formed individual to the consecutive new one to
be formed. The chain of these numbers forms in itself an
exceedingly useful instrument for the human mind; it pre-
sents an inexhaustible wealth of remarkable laws obtained
by the introduction of the four fundamental operations of
arithmetic.

These long sentences suggest a powerful mind moving steadily under a full head of steam; but words such as *act, creation, infinite, inexhaustible,* and *arising* indicate something more curious and even rhapsodic than the otherwise sober concerns of a provincial German pedagogue, a scheme at work, and so a vision, something primitive, powerful, and unexpected.

Counting is at the heart of Dedekind's meditation, a purely mental undertaking, something that the mind *does* or that human beings *do*; and it is counting in turn that engenders the "successive creation" of the natural numbers. To those proposing to explain counting in other terms, Dedekind offers no support. *Counting is primitive.* It is what it is, and not some other thing; it cannot be further analyzed.

Having accepted counting as given, Dedekind also accepted in zero an initial number and so a beginning, something hard, ineradicable and indubitable, the number zero functioning very much like the cosmic egg that Father Georges Lemaître discerned when he first entertained the hypothesis that the universe had its origins in a Big Bang.

The creation of the natural numbers proceeds from the action of succession on the number zero:

> *From zero, one.*
> *From one, two.*
> *From two, three.*
> *From three, four . . .*

The result emerges in stages, very much like a tower rising where no tower might ever have been expected.

The effect is eerie, because the numbers are wonderfully various, different in their properties, riotous, even numbers against odd, perfect numbers against imperfect numbers, square numbers, abundant numbers, deficient numbers, Mersenne primes, prime numbers against all the rest, numbers that are small, and those that are large. No physical tower *ever* arises in this way.*

The reduction of experience encompassed by this idea is far more radical than comparable claims made about the fundamental particles in physics.

STARTING AT ZERO, ADDING BY ONE

Succession makes for a very simple representation of the natural numbers. For any number x, the successor of x is designated by $S(x)$, so that $S(0)$ is one, and $S(1)$ is two, and $S(2)$ is three.

There is no need to write things out by hand. Succession may itself be abbreviated by iteration, with $SS(0)$ designating the successor of the successor of zero.

Or indicated by arrows:

$$0 \rightarrow S(0) \rightarrow SS(0) \rightarrow SSS(0) \rightarrow SSSS(0) \rightarrow$$

Although succession cannot be defined, it may be replaced by a more familiar operation. The successor to a given number is that number plus one. This is what succession has always meant, but giving stress to plus *one* is a way of enhancing its steplike nature.

Start at zero. Keep going up by one.

---

* These distinctions belong to the zoology of number theory. The even numbers are those that may be divided by two; not so the odd numbers. A prime number is divisible by itself and the number one, and nothing else. A perfect number is equal to the sum of its divisors. Thus the number six is perfect, because six equals three plus two plus one. A square number is a number that may be expressed as the square of another number. Twenty-five is an example. What else? An abundant number is one that, like twelve, is less than half the sum of its divisors when its divisors include the number itself. Thus twelve is less than half of twelve plus six plus four plus three plus one. Deficient numbers go the other way. A Mersenne number is a number one less than a power of two. Seven is a Mersenne number since it is one less than two cubed. A Mersenne prime is a prime Mersenne number.

There is then

$$0 \to (0 + 1) \to (0 + 1) + 1 \to ((0 + 1) + 1) + 1 \to \ldots$$

$S(0)$ is, after all, quite the same thing as 1, and $0 \to S(0)$ the same as $0 \to 1$, and $0 \to 1$ the same—*no?*—as $0 \to 0 + 1$.

The idea that the natural numbers arise by means of *addition by one* is welcome. The next number up? It is that number plus one. The next number up is the next number plus one.

It is what everyone says.

Me too.

Nonetheless, mention thus of addition seems to suggest that a concept supposed to appear by the front door has, by an improper maneuver, been introduced at the rear.

Adding by one is *adding* by something, and adding by *anything* is a concept that has not yet been defined by, or even mentioned in, the Peano axioms.

Although often made, this objection has no enduring vitality. It comes to nothing. What is at issue in adding by one is not a new concept but a new convention. Instead of $S(0)$, there is $0 + 1$. The number 3,642 is just the number 3,641 plus one, and it is also the successor of 3,641, or $S(3,641)$. There is no new way of proceeding, and only the old way of having proceeded.

But if adding by one adds nothing to the content of the Peano axioms, it does add something to their character, the way they reflect light. It reveals an unsuspected symmetry, one in which *adding by one* is matched nicely to *starting at zero*.

The numbers zero and one, Gottfried Leibniz speculated in the seventeenth century, are the touchstones of creation itself, the universe—yes, the *universe*—forged from the conflict between two numbers, the first representing nothingness, the second being.

. . .

DEATH DEFERRED

Richard Dedekind's long career as a mathematician began when in 1854, the great Gauss approved his dissertation, the remote, unapproachable old man observing only that he found Dedekind's work "satisfying," and ended in 1916, when Dedekind surprised the mathematical community by dying long after he was commonly thought dead. Dedekind was born in 1831 in the village of Braunschweig, and, like that of a number of other German mathematicians of the mid-nineteenth century, his early upbringing embodied the positive aspects of a culture that was Protestant, pious, pedantic, and patriotic.

There is little in Dedekind's early education to suggest a mathematical personality. Braunschweig was not a center of mathematical culture, and the education that Dedekind received at the Collegium Carolinum was straightforward and of that time and that place.

It was not widely remarked that his genius was electrifying. Dedekind had come to mathematics by means of a detour. Originally interested in physics and chemistry, he had discovered in these subjects an attitude alien to his own. The chemists could give no very lucid account of what they were doing, and emerged from their laboratories evil-smelling, their hands burned by various acids and bases. They formed a priesthood of hard, gifted, but practical men. The great physicists of the nineteenth century, on the other hand, were visionaries, and in classical physics, they created a subject like nothing ever seen before. When, in the twentieth century, relativity and quantum mechanics shattered this cathedral in thought, Wolfgang Pauli could look back and, with the eyes of someone who had seen glory fade, recognize its power. But in their attitude toward pure mathematics, nineteenth-century physicists were as practical as the chemists. They were ambitious to get to where they were going. How they got there was not their most pressing concern.

By nature fastidious, Dedekind withdrew from the physical sciences.

If his life had lacked drama before his emergence as a mathematician, it lacked drama afterward. Intellectual ostentation he never sought and rarely indulged. Dedekind formed a number of close mathematical friendships, especially with Peter Dirichlet; he engaged in research; for a time he taught at the Zurich Polytechnic. His views were calculated to bring him into conflict with Leopold Kronecker, for Dedekind was a remarkably generous mathematician, prepared, as Kronecker was not, to follow mad men into bad lands. He was a friend to Georg Cantor and admired his work, and yet curiously enough, while Kronecker did everything in his power to undermine Cantor, and to impede the development of his theory, there is little to suggest that he was provoked by Dedekind to any political unkindness, one reason, I suppose, that Dedekind always seemed a man at ease with himself, serene by nature and by the grace of circumstances.

DOWN WITH EUCLID

The idea that the natural numbers are generated by an act of counting is now deeply entrenched in a general human consciousness.

The entrenchment leads to what seems a persuasive and natural way of describing the applications of **AEM**. It is counting numbers that leads to counting things, the sheepherder's one *sheep*, two *sheep*, three *sheep* a pale, polluted reflection of the mathematician's primordial *one, two, three*.

It is not only in counting sheep that counting counts. We now think of distance as a derived number. Moscow is *one thousand* miles from Prague. A tennis player is *three* points from victory. And the victim, poor thing, was *moments* away from death. A pity that she clung to her purse instead of her life. The numerical measure comes *before* the physical one, three points from victory three *points* in virtue of the rules of tennis, but *three* points in virtue of the prior progression *one, two, three.*

If natural, this view is also new in the long arc of mathematical history. Had Dedekind appeared among ancient Greek geometers, his rumpled toga hitched, they would have heard him out with a hard, discouraging skepticism, a sense that he had gotten things backward. From *their* point of view, counting is the subordinate activity, and *distance* the fundamental idea.

An old form of experience is at work, older by far than Euclid, one embedded in the double observation that things move and time passes. Things in nature move, thus creating spatial extent; time passes in consciousness, thus temporal extent. Distinct objects of perception—*here and there, now and then*—invite the mind to fill what is between them with something that enforces their separation. Were that thing not there, points would collapse on one another and we would be returned to the terrible, all-encompassing place without difference in time or space.

These are the ideas that Greek geometers refined. The pinhead of a moving point, they argued, determines a line in the plane; the sail of a moving line, a surface; and the fold of a moving plane, a volume. A *point* in Greek geometry is an entity—there *it* is, after all—but one with *no* spatial extension. A point, Euclid emphasizes in his definitions, is "that which has no parts." The partlessness of points implies their lack of extent. A point divided into two parts would, after all, be two parts in extent, two extents in length, or in area, or in volume, but, whatever the case, there would be *two* of them.

It is in Books V and IX of the *Elements* that Euclid turns to arithmetic, and the derivation of numbers.

The *here and there, now and then* are in Euclid's system both subordinated to the idea of progressively increasing distances. In moving from the place it was, a point determines a line segment. The sequence of segments, as they inch forward, is the Euclidean analogy of the way, within **AEM**, the numbers are created by counting. Euclid thus assigned to line *segments* some properties of the numbers themselves. The result is a congeneric arithmetic, one racially similar to the real thing, but strange. So the very first sentence of Book VII affirms that "A *unit* is that by virtue of which each of the things that exist is called one." A "number," Euclid goes on to say, is a "multitude of units." The scheme is one in which a given line segment is chosen arbitrarily, line segments beyond compared with the unit and compared with one another.

This way of thinking and seeing is alien to **AEM,** because **AEM** begins with the primitive act of counting and so begins with a first number. It has nothing to do with points, lines, volumes, or whatever may be made from them. It is indifferent to geometry and is its own thing. Every mathematician who has contributed to **AEM** has argued that in the purity of its conception it is superior to geometry.

"Death to triangles," the French mathematician Jean Dieudonné urged.

"Down with Euclid."

8

*Addition is one of the four operations of **AEM**. The others are multiplication, subtraction, and division. In each
operation, two numbers are taken to, or yield, a third
number. There is two and there is three—two numbers;
and there is 2 + 3, whereupon there is a third number,
in five.*

ADDITION

But if addition takes numbers to numbers, it cannot do so literally. Numbers cannot be taken anywhere. They do not yield anything, because they can no more yield than resist. A metaphor is
at work even in so simple a declaration as two plus three *makes*
five. If *making, taking,* and *yielding* unavoidably remind us of the
purely human agents doing the making, taking, or yielding, the
word "operation" correspondingly suggests an urgent flight into
abstraction on their part, an antipode to earthiness and so a sign
of mathematicians' unwillingness to repose their confidence in
any physical activity. But even when the operation of addition is
purged of its association with the world of action, the ideas that
result retain something of an ancestral physical memory.

Merchants and mathematicians have long designated the
sum of two and three by placing a summation sign between their
names: 2 + 3. The symbol + has the very considerable advantage that by extending its arms in fellowship, it exhibits what it
denotes, a binding of numbers. The sign of the cross has a genius
in mathematics as well as Christianity. The symbols 2 + 3 could
also be written as +(2,3), and, if written in this way, then written as
$f(2,3)$. Such is the mathematician's functional notation, the sym-

bol *f* denoting a function, a mapping or an operation. If 2 + 3 = 5 says that two plus three equals five, then so does *f*(2,3) = 5. This function takes two numbers to a third; the two numbers make up its argument, the third number its value, and functions are generically instruments taking arguments to values.

Functional notation, if unfamiliar, offers a twofold advantage over tradition. There are within mathematics a great many functions, and it would be tedious to find a new symbol for each of them. The polyvalent symbol *f*, understood now as addition and then as multiplication, is sleek, elegant, and handy. If different functions must be distinguished, there are always new symbols in *g* and beyond *g, h.*

In symbolizing 2 + 3 = 5 as *f*(2,3) = 5, the mathematician has created a visual tableau that, by means of symbols leaning on symbols, suggests the hidden urgency of action that underlines *all* mathematical operations. Even in mathematics, something is done only when someone has gotten something done. Numbers do not add themselves.

The operations of **AEM** fall naturally into groups. Addition and multiplication are alike because, with the exception of multiplication or addition by zero, they inevitably progress upward. The sum of three and five is greater than three or five, and so is their product. Addition and multiplication also have a curious emotional affinity. Although exceptions are obvious, the operations of addition and multiplication very often serve sunny ends; *The more the merrier* and *Be fruitful and multiply* both express the human instinct to flee from nothing.

On the other hand, addition is linked naturally to subtraction, because subtraction undoes addition. If five plus three is eight, then eight minus three is five. A sense of symmetry is at work. No culture, I suspect, capable of adding five bricks (seashells, pebbles, goats) to three of them and getting eight as their sum has ever failed to recognize that if bricks could be added to bricks, bricks could as well be withdrawn from bricks.

For the same reason, multiplication and division are affably associated. Five times three is fifteen, and fifteen divided by five is three.

Within the context of the natural numbers, the elective affinities between addition and subtraction and multiplication and division are notoriously unstable. The result, after all, of subtracting ten from six is nothing at all, and so is the result of dividing three by two. If, as adverted, the Undoer of Subtraction is symmetrically associated with the Doer of Addition, just why should Undoing flicker out inappropriately when five is taken from three and not when three is taken from five?

Subtraction and division *are* made whole in elementary mathematics, the operations restored to symmetry. New numbers are required in the form of the fractions and the negative numbers. If these new numbers restore the symmetry of **AEM**, they also compromise its purity. Neither the negative numbers nor the fractions could ever be considered a gift from God.

Mastering the operations of **AEM** is one of childhood's tasks. Children are taught simple sums by rote; they are trained to calculate complicated sums by reducing them to simple sums. Whether sums are simple or complex, children are taught them without ever being taught what the operation of addition might mean.

Until the late nineteenth century, mathematicians did no better. If men of genius such as Gauss, Galois, or Abel failed to provide a definition of addition, it was because they failed to see that one was required. When mathematicians did provide a definition, they failed again to see that their definition required a justification; and when, fifty years later or so, mathematicians did provide the appropriate justification, one of the simplest and most evident operations undertaken by the human mind—adding two numbers—acquired a degree of conceptual richness that no one, mathematician *or* merchant, had ever suspected that it possessed.

Richard Dedekind's literary voice conveys a sense of the drama involved. It is unusual in the literature of mathematical memoirs in conveying both self-assurance and self-awareness; it is the voice of a man at ease with himself, and well disposed toward his readers.

And it is avuncular. His ideas, Dedekind writes smoothly, "can be appreciated by anyone possessing what is usually called good common sense; no technical, philosophic or mathematical knowledge is in the least degree required."

But Dedekind had struggled with his own ideas, and he knew that they were both original and subtle, and so difficult to grasp.

In the end, Dedekind defers to the facts by admitting them.

"Many readers," he writes, "will scarcely recognize in the shadowy forms which I bring before him his numbers, which all his life have accompanied him as faithful and familiar friends."

9

The words "the definition of addition" may suggest that having deferred the matter for centuries, the contemporary mathematician is now in a position to say what addition means once and for all.

Not so.

BY DESCENT

The definition of addition is a way of calculating the sum of two numbers in a finite series of steps. It is not a proper definition in which the *definiendum* vanishes in favor of the *definiens*. It is, instead, a recipe for adding numbers, a way of proceeding. Because the recipe requires *only* a finite series of steps, it inevitably comes to an end. This is logically reassuring. If the recipe requires only a finite series of steps, it is also true that it works only to a finite extent. This is less reassuring.

Definition by descent, as its name implies, exploits the expanding tower of the natural numbers in the most obvious of ways. Two steps are involved. In the first, the sum of two numbers is referred downward to the sum of two smaller numbers; and in the second, the difference in sums is made up by succession, or addition by one. The sum of four and three is the sum of four and two *plus one.*

There is in all this an unavoidable suggestion of contamination, as if the essential thing, and that is the definition of addition, were forever being avoided, and avoided precisely in stages or steps. If in order to understand the sum of four and three it is necessary first to understand the sum of four and two, how has an understanding of *addition* been improved? Is this not a little like defining a brother as a male sibling and then observing that a male sibling is a brother?

The charge of circularity in definition by descent is understandable, but mistaken. In defining the sum of four and three in terms of the sum of four and two, the mathematician *has* violated an ancient logical protocol. A proper definition must *eliminate* whatever it is that is being defined.

Yet, if the definition of addition involves something akin to a circularity in thought, the defect is discharged by descent so that in the end it disappears.

The sum of four and three is defined by the sum of four and two plus one.

But going down still further, the sum of four and two is *then* defined by the sum of four and one plus one, and then, ultimately, by the sum of four and zero plus one taken twice.

With three gone from all further calculations, the number four follows into the void. Addition vanishes as an operation. The only thing left is counting—pure act. The sum of four and three is zero plus one plus one plus one plus one plus one plus one plus one.

This idea is so simple and so persuasive that often it is the first thing taught to children. To determine how many blocks there are when four of them are added to three of them, the blocks are pushed together and then counted.

This is what definition by descent accomplishes.

Sort of.

NOVEL NOTATION

The definition of addition must by a verbal maneuver give content to the sum of two numbers, and it must do so for *every* two numbers x and y. The symbols $x + y$ have some of the false familiarity of foreign words imperfectly understood (*étiquette, coup d'état, Weltanschauung*). Within $x + y$ the symbol + stands in place like a martyred dwarf, one extending its arms in fellowship toward two *letters* in *x* and *y*. The symbol + means plus; but the letters *x* and *y* cannot trigger the addition of anything, because they do not obviously designate numbers, an observation that at once suggests the limits of familiarity.

The point is not foolish; the notation does *not* explain itself.

The requisite devices entered mathematics more than a thousand years ago, another oblation of the Arab Renaissance, but their refinement was the work of the nineteenth century, Giusseppe Peano, Gottlob Frege, C. S. Peirce, George Boole, Augustus De Morgan, and later Bertrand Russell, comprising a regiment of notational champions, like the Forty at Thebes.

Symbolized by late alphabetic letters such as *x, y,* and *z,* variables perform some of the offices of ordinary pronouns in English, the Anglo-Saxon *I, you, he, she, it, we,* and *they* permitting a form of indefinite reference, with *I came, I saw, I conquered* leaving undeclared just *who* came, *who* saw, and *who* conquered *what.*

We know quite well that *it* was Julius Caesar, the author and the subject of his own remarks, but we know what we know from context and not syntax; and, what is more, just *who* is this *we?*

While English pronouns are serviceable in describing who is doing what, they are often awkward in handling divided references, with *he thought he might as well go* leaving it open whether the pronoun *he* refers to one man or two.*

There is in ordinary language a fatal tendency toward ambiguity, one that in mathematics must be controlled. Where would anyone be if *x* is greater than *x* had some of the lingering uncertainty of *he thought he might as well go?*

Variables are of the family of names, and so are symbolic forms. The distinction between symbols and what they name is in force.

---

* *John thought that he should go* is ambiguous; but *he thought that John should go* is not. Curious.

The letter x is a symbol, a bit of linguistic debris, something entering **AEM** from the back of the alphabet, but in the statement that x is greater than five, the variable x is doing work in the real world. As *their* name suggests, variables are variable in what they designate. The variables x, y, z are intended to designate the numbers; this is their domain of application. The symbols $6x$ signify, by virtue of their proximity, the forthcoming marriage in multiplication of six and *some* number x; the symbols $x > 6$ that *some* number x is greater than six; and the symbols $x > y$ that *some* number x is greater than *some* number y.

Which number? Unlike the symbol 6, the symbol x in $x > 6$ is not occupied in picking out a particular number. The force of $x > 6$ is that x designates *some* number or *other*, the number's identity not specified beyond the fact that it is greater than six. This form of indefinite reference makes possible the governing technique of the modern algebraic theory of equations, the method of indirect identification, as when x is identified simply by the fact that when squared it is twenty-five.

Like lawyers, logicians have a notation beyond what laymen need, with *there exists* represented by a backward-facing $\exists$, a quantifier that with its three outstretched arms seems to be trying hopefully to reach the sources of creation itself.

Variables and quantifier acting together serve to impart a compressive force to what would otherwise be tedious in English; *there is a number such that when it is squared the result is twenty-five* disappears entirely in favor of $\exists x \, (x^2 = 25)$.

Neat, elegant, brisk, true, and *short*.

THREE CLAUSES

The conceptual problem never faced in childhood now emerges in its full generality:

How to define $x + y$ for *any* two numbers x and y, given only succession or $0 + 1$?

Three clauses are required. The first establishes zero as a number that in addition does nothing. For every number x:

$$1 \qquad x + 0 = x.$$

The second clause in the definition of addition is a reminder. What must be defined and so given meaning is the sum $x + z$ of any two natural numbers x and z—four and three, say.

Whereupon the obvious: For *any* number z beyond zero, there is *always* some number y such that $z = y + 1$. Four is, after all, three plus one; three is two plus one; and two is one plus one. There is no reason to suppose that this principle fails anywhere, and, indeed, it may easily be derived from the Peano axioms.

The second clause in the definition of addition thus affirms for *any* numbers x and z and for *some* number y that:

$$2 \qquad x + z = x + (y + 1).$$

It is the third clause in the definition that expresses the principle of definitional descent. For any numbers x and z, with the guarantee (by 2) of some appropriate y,

$$3 \qquad x + z = x + (y + 1) = (x + y) + 1.$$

The clause is read from left to right, the number z vanishing in favor of the number $y + 1$, and the sum of $x + (y + 1)$ then vanishing in favor of the sum of $(x + y) + 1$.

It is very strange and very moving that three short symbolic clauses should somehow capture the flow of addition, little numbers added to little numbers, or big numbers to even bigger numbers, little to big, big to little, an obviously endless number of combinations disciplined by language and subordinated to inference.

Peevish? Are these exercises in symbolism *peevish*? Or are they instruments of some considerable act of intellectual daring?

The latter; the latter *entirely*.

IN SPACE AND TIME

A recipe has been given, a procedure specified. Who is doing *what* to *whom*? It does not hurt to ask.

It is $x + (y + 1)$ that is processed; it is $(x + y) + 1$ that is processing. These expressions are different not in the numbers that they designate, since they designate the same number, but in the way in which the number is designated. The number $x + (y + 1)$ is designated by *first* adding 1 to the number y, and *then* adding the result to the number x. The number $(x + y) + 1$ is designated by *first* adding the number x to the number y and *then* adding 1 to the number $(x + y)$.

Parentheses are now giving the stage directions. Scholastic logicians called parentheses and the like syncategorematic. Matched in pairs, parentheses serve to indicate associations within mathematical expressions and so play a crucial role in canceling ambiguities. It hardly matters when it comes to $2 + 3 + 5$ whether 2 goes with 3, or 3 with 5. It all comes out the same. But in $2 + 3 \times 5$, it matters a great deal: read as $(2 + 3) \times 5$, the result is twenty-five, and read as $2 + (3 \times 5)$, it is seventeen.

In a far deeper and far more mysterious sense, parentheses in mathematics are temporal markers, the associations that they indicate correlated to things that we do, doubled curves of the parentheses marking not only an association of numbers but a beginning and an end.

Mathematicians very often write that the natural numbers are beyond space and time. In some sense, this is obviously so, although what *beyond* might mean in this context is not clear. *How far* beyond? If this is casually to suggest a view that approaches parody, its converse does no better. What might it mean to say that like every other thing, the natural numbers appeared at a certain date, and that after a number of years they might disappear? It is surely incoherent, because any attempt to fix the appearance of the natural numbers in a temporal stream must assume their existence in order to describe the stream. So perhaps the natural numbers are beyond space and time in some sense of *beyond*, some sense of *space*, and some sense of *time*, these involving no commitments whatsoever to anything definite.

That seems safe enough to me.

What remains is this. Whatever the natural numbers may be, or the residence they may occupy, the elementary *operations* such as addition remind us, by means of the curve of their parentheses, that in adding two numbers together something is done *now*, and some other thing done *later*.

"What we cannot comprehend within space and time," the physicist Erwin Schrödinger observed, "we cannot comprehend at all."

FOUR PLUS THREE

The definition of addition is now ready to do work in the real world by determining the sum of 4 and 3.

If for any number 2 there is an equivalent number $y + 1$, then, in the case of the number 3, the requisite $y + 1$ is simply $2 + 1$. Where before the number 3 loomed large, in what follows it is designated from below by two plus one. The appropriate substitution is evident in the chain of inferences that follow:

$$4 + 3 = 4 + (2 + 1).$$

Now

$$4 + (2 + 1) = (4 + 2) + 1$$

by the third clause in the definition of addition, the definition taking hold of the parentheses in 4 + (2 + 1) and forcibly moving them to the left in (4 + 2) + 1. The magic by which 3 gave way to 2 + 1 works as well with 2, which gives way to 1 + 1.

So

$$(4 + 2) + 1 = (4 + (1 + 1)) + 1.$$

That third clause, invoked once again, yields

$$(4 + (1 + 1)) + 1 = (4 + 1) + 1 + 1.$$

Replacing 1 by 0 + 1,

$$(4 + 1) + 1 + 1 = (4 + (0 + 1)) + 1 + 1,$$

so that by invoking the third clause again,

$$(4 + (0 + 1)) + 1 + 1 = (4 + 0) + 1 + 1 + 1.$$

But 4 + 0 is just 4, by the first clause in the definition, doing something useful for the first time. It follows that

$$(4 + 0) + 1 + 1 + 1 = 4 + 1 + 1 + 1.$$

On returning the tail of this inference to its head, we see that

$$4 + 3 = 4 + 1 + 1 + 1.$$

Purists may wish to purge four from this account, expressing everything in terms of zero and its successors. The result is 0 + 1 + 1 + 1 + 1 + 1 + 1 + 1. This austere account may be made more austere still when even plus one is purged in favor of succession itself. The sum of four and three is nothing more than the seven-fold successor of zero: SSSSSSS(0). In this way, a number conceived in mystery—zero—is assigned a generative power entirely at odds with its role in signifying nothing.

Whatever the notation, addition has vanished in favor of the "simplest arithmetic act" that Dedekind invoked.

Dedekind's calm animadversions are now pertinent—as they have always been. Asking a first-grade child—Daphne, as it happens, in my care for a midge-filled summer afternoon—for the sum of four and three is already to push her beyond her computational limits, as I discover on posing the question and seeing her chapped lips turn downward in a drooping pout. But counting by one is not, as, with a good deal of recovered enthusiasm, and a fine spray of saliva, she proceeds to do just that, reaching seven in a blaze of glory and proceeding onward very happily.

And yet counting by ones to seven and adding four and three are the same thing.

They are *precisely* the same thing. We are now among the shadowy forms of which Dedekind spoke. The first great mathematical masterpiece of concision and intellectual supremacy in **AEM** has been revealed.

10

Having discovered how numbers might be added, merchants in the ancient world certainly knew how they might be multiplied. Their techniques were a part of the Sumerian empire's scribal art.

MULTIPLICATION

They stretch back to the beginning of recorded history. This might suggest that both addition and multiplication represent a gathering inevitability in the progression of the human mind: to *have* thought of addition is to think *next* of multiplication.

Like addition, multiplication is defined entirely within the ambit of the natural numbers. Operations do not come more naturally than these. Subtraction and division, to be sure, may be given a definition of sorts using nothing more than zero and the numbers that follow, but the definition reveals only mutilated operations, ten minus five doing well, but five minus ten crippled from the first. And the same thing is true of twelve divided by two. It is fine. But two divided by twelve requires resources that the natural numbers do not provide.

If addition and multiplication *both* appear as operations in any culture sophisticated enough to have a sense of the numbers themselves, it might seem obvious that they must be distinct. Why would sensible Sumerian merchants bother themselves with multiplication if, in the end, addition does as well, or at least does as much?

If the point seems obvious, it is not because it is widely endorsed in elementary education. On the contrary. Multiplication is defined in the textbooks as repeated addition. Five *times* six, or 5 × 6 (or 5 · 6), is just six taken five times (or five taken six times):

$$6 + 6 + 6 + 6 + 6.$$

This interpretation is said to reveal an economy of effect. A sixfold addition requires six acts, a one-time multiplication only one. Thus a definition: the product of any two numbers x and y is y taken or added to itself x times.

While obviously correct, this definition is not obviously satisfying. For one thing, in defining the product of five and six as six taken five times, it reintroduces a concept with at least a linguistic link to multiplication itself.

Five *times*?

For another thing, if five times six is defined as six taken five times, is six taken five times the same thing as five taken six times? The definition leaves open the alarming possibility that it may not be.

It does not say.

The definition also gutters out inconclusively when a request is made for the product of six and zero. Just what *is* zero taken six times? Zero, I suppose. But, by parity of reasoning, shouldn't six taken zero times still be six? How else would one understand the idea of taking a number no times at all?

And, finally, if a reduction of multiplication to addition is at work, as the textbooks argue, why is 6 *plus* 0 equal to 6, but 6 *times* 0 equal to 0?

Questions such as these suggest that, when it comes to the details, multiplication is an invitation to incoherence.

Sumerian merchants were quite correct. Multiplication is an operation that is distinct from addition.

It is its own thing and answers to different constraints.

THE DEFINITION OF MULTIPLICATION

The definition of multiplication is, like the definition of addition, an exercise in definitional descent. The definition trades, moreover, on the assumption that addition has already been defined. This satisfies the intuition that among the natural numbers addition is primary. The definition contains three clauses.

The first establishes zero as a number that in multiplication returns to itself. For every number x

$$1 \qquad\qquad x \cdot 0 = 0.$$

The contrast between comparable clauses in the definition of addition and multiplication is striking. In addition, zero returns the number; in multiplication, it returns *zero*.

The number one, on the other hand, does for multiplication what zero does for addition, since for any number x, $1x$ is simply x. Both zero and one are for this reason called the identity elements of their respective operations.

The second clause in the definition of multiplication functions as a reminder. For any number z beyond zero, there is always some number y such that

$$2 \qquad\qquad z = y + 1.$$

Having been introduced in the definition of addition, clause 2 reinforces the sense that the numbers form a progression, increasing at each step by one.

The third and final clause in the definition of multiplication triggers definitional descent. What must be defined, and so given meaning, is the product of any two natural numbers x and z. This the clause accomplishes by the same technique employed in the definition of addition:

3 $$x \cdot z = x \cdot (y + 1) = (x \cdot y) + x.$$

Going from left to right, clause 3 first authorizes the dismissal of $x \cdot z$ in favor of $x \cdot (y + 1)$.

And then the dismissal of $x \cdot (y + 1)$ in favor of $(x \cdot y) + x$. Weight shifts inexorably from the right.

If definition by descent is common to addition and multiplication, it proceeds by radically different means. The definition of addition works by *association*, so that $x + (y + 1)$ is recast as $(x + y) + 1$. The definition of multiplication works by *distribution*, with x in $x \cdot (y + 1)$ exerting multiplicative force over both y and 1, the needed recasting yielding $xy + x$, and *not* $xy + 1$.*

Whatever its other merits, clause 3 has the great virtue of consorting companionably with common sense. The product of 5 and 4 really is 5 times $(3 + 1)$, which really is $(5 \text{ times } 3) + 5$, which in turn really is what it is supposed to be, and that is 20. All traces of addition may ultimately be purged in descent; but in multiplication too, as when $(5 \text{ times } 3)$ is replaced by $(5 \text{ times } 2)$ and then $(5 \text{ times } 1)$ and ultimately by $(5 \text{ times } 0)$. And this, as the very first clause in the definition reminds us, is nothing at all.

This is another way of saying that the definition *works*, never a bad thing in mathematics, or anywhere else.

It works, moreover, to various other good effects, and so comprises a gift that keeps on giving. Given the definition, the fact that the number one is an identity element for multiplication follows as a direct inference.

The proof occupies a single line: For any number x, $x1 = xS(0) = (x0) + x = x$.

. . .

* Association and distribution are terms of art. They require an explanation. One is forthcoming in Chapter 14. These remarks are a coming attraction.

THE PRODUCT OF THREE AND TWO

And here is the definition of multiplication at work in determining the product of three and two.

The number 2 is plainly $1 + 1$, and so

$$3 \cdot (1 + 1) = (3 \cdot 1) + 3.$$

But $3 \cdot 1$ is $3 \cdot (0 + 1) + 3$, and $3 \cdot (0 + 1)$ is $3 \cdot 0 + 3$, in which case $(3 \cdot 1) + 3$ collapses into

$$3 + 3,$$

with multiplication as a separate operation displaced in favor of addition.

On returning $3 + 3$ to its head, we see that

$$3 \cdot 2 = 3 + 3.$$

It is easy enough to carry out computations until *all* numbers disappear into the staccato of succession from zero. This is an exercise too familiar to require doing again. What does merit appreciation is the process: Multiplication in favor of addition and multiplication, but lower down. Addition in favor of adding by one. Multiplication in favor of zero. Adding by one in favor of finger counting, and finger counting in favor of the beating human heart, the place and sound where absolutely elementary mathematics begins.

EXPONENTIATION

Addition and multiplication have now dwindled and disappeared.

There is next exponentiation. The expression 10^n designates ten undergoing a form of auto-intoxication. Ten squared is ten times itself; ten cubed, ten times itself times itself; and so up to 10^n, which is ten times itself n times.

The number ten is the base in this affair; the number n, its exponent. The conjunction of base and exponent determines a new number, and so exponentiation is functionlike in its effect. It does something. Unlike addition and multiplication, exponentiation does not have its own symbol, its effect conveyed by the position of the exponent over the head of its base.

In defining exponentiation, any allegiance to the number ten may be dismissed. That allegiance was forged with familiarity in mind. There is no point in demanding that exponentiation be dominated by a single number. *Any* natural number can serve as the base of the exponential function, 2^{17} as eager to multiply itself by itself as 10^{24}.

Definitional descent is required to bring order to exponentiation. Three clauses. The first conveys the fact that exponentiation to zero returns any number to 1. For any x greater than 0,

1
$$x^0 = 1.$$

The second re-re-affirms: any natural number z may be expressed in terms of its predecessor y, so that

2
$$z = y + 1.$$

And finally descent:

3
$$x^{y+1} = (x^y) \cdot x.$$

With descent, a return to the familiar. The number seven raised to the third power is seven times seven times seven, and seven raised to the third power is simply seven raised to the *second* power times seven.

If the consequences of descent are familiar, not so its very first clause, which prompts the question why 10^0 should be one? There is something about the symbol 10^0 that suggests one of those horrible flounders with eyes on the same side of its head. Never mind the number ten then. How is it that 3^0 is one? Three raised to one is three. It should not be more and it cannot be less. But three raised to *zero*? Why is it one? Is there not a wandering point of incoherence in saying that three *times* zero is zero but that three *raised* to zero is one?

This is what students always ask. And they are right to ask it.

Facts do in the end cohere, for suppose that instead of saying that 3^0 is one, we were to say that 3^0 is zero.

In that case, what of 3^1?

The number one is equal to the number $0 + 1$, and so 3^1 is equal to 3^{0+1}.

From which it follows that 3^{0+1} equals $3^0\,3^1$.

Whence $3^0\,3^1$ must be zero if 3^0 is zero.

So, if $3^0\,3^1$ is zero, then 3^1 *must* be zero too.

This argument would suggest that exponentiation is an operation that returns everything to nothing.

These are not encouraging conclusions. Do they amount to a proof of the contrary? Are they unassailable?

Not quite. They establish only that *if* definitional descent captures something essential about exponentiation, then x^0 *cannot* be zero.

They do not establish that it *must* be one.

But just between us, what else could it be?

EXPONENTIAL POWER

The definition of exponentiation serves to deliver, by means of easy steps, a bouquet of exponential powers.

The exponentiation of the number one by any number whatsoever is always one. Such is the force exerted by the gravitational field associated to the number one. In symbols: $1^z = 1$.

If exponentiation respects a certain local gravitational field in the case of the number one, with respect to other numbers it has a way of violating established parenthetical boundaries. When a number x is raised to a certain power y, and the result then raised yet again to the power z, the mathematician would ordinarily take $(x^y)^z$ to designate a two-step operation: up goes x to x^y, and then up goes x^y to $(x^y)^z$. As it happens, exponents can enter cohabitation across parenthetical lines. The two steps are not necessary: $(x^y)^z = x^{y \cdot z}$.

By the same token, an exponent, when applied to the product of two numbers, can be assigned to both numbers, the burden of multiplication divided: $(x \cdot y)^z = x^z \cdot y^z$.

There is nothing in this that might prompt loitering or lingering. Not so what remains, an all-important exponential identity— *the* all-important exponential identity—one by which fortunes were made, empires created, and the scientific revolution of the seventeenth century given a very considerable boost.

The identity is simple enough: $x^y x^z = x^{y+z}$. In multiplying 10^3 by 10^5, it is not necessary to go through two exercises in exponentiation and then multiply the results. It suffices to *add* exponents and then carry out exponentiation *once*: $10^3 \cdot 10^5$ is 10^8.

Now, exponentiation is an undertaking that proceeds upward from a base to an exponent. What goes up must come down. Hence the logarithm, the idea that John Napier introduced in the seventeenth century in his *Mirifici Logarithmorum Canonis Descriptio (A Description of the Marvelous Canon of Logarithms)*. The logarithm of a number is again a number, one determined by going down after going up. Going up yields one hundred from ten and the exponent two. Going down yields two from one hundred and the number (or base) ten. The number two is the logarithm of one hundred. And the number one hundred is the anti-logarithm of the number two.

These gathering algebraic forces allowed the mathematician to reduce multiplication to addition, and so to evade the computational complexity of one operation in favor of the computational simplicity of another.

A very simple inference governs this form of mathematical alchemy, all glitter and glow.

For any two numbers x and y

$$xy = 10^{\log xy}$$

but then

$$10^{\log xy} = 10^{\log x + \log y}$$

and so

$$10^{\log x + \log y} = xy.$$

The equality expressed by $10^{\log xy} = 10^{\log x + \log y}$ is the flow and overflow in action of that all-important exponential identity.

Flow: $10^{\log x}$ times $10^{\log y}$ is equal to $10^{\log (xy)}$ since both sides of this equality are equal to xy.

Overflow: $10^{\log x}$ times $10^{\log y}$ is $10^{\log x + \log y}$ because $x^y x^z = x^{y+z}$.

Logarithms and anti-logarithms may be calculated in very fine detail. In the tables that result, physicists and engineers, geologists, navigators, and men who understood how steam expands had for the first time a remarkably efficient calculating tool at their disposal, exponents demonstrating the unheard-of power to scramble out of what they had been scribbled into.

11

*Positional notation is the principle by which numbers are named, but its use has thus far been limited to numbers designated by only two names—twenty-seven, say, or thirty-two—numbers, that is, of the form **ab**.*

THE GREAT DICTIONARY

The introduction of addition, multiplication, and exponentiation provide for a system by which all of the numbers can be named—not all at once, perhaps, but systematically, by means of a recipe or rule. The result is the Great Dictionary of the Natural Numbers. The dictionary is great both as a mathematical achievement and as one of the markers of civilization itself, for without the dictionary there would be no such thing as Western science, no way to capture in symbols the ever-expanding tower of the natural numbers.

The Great Dictionary of the Natural Numbers is designed to accommodate a twofold form of transit. Read in one way, entries allow the mathematician to go from the names of the natural numbers to the numbers themselves; read in reverse, those same entries allow the mathematician to go from the numbers back to their names.

The coordinating link between names and numbers in the Great Dictionary is the concept of a base, a natural number by which all other natural numbers may be expressed. In the most current version of the Great Dictionary, it is the number ten that serves as the system's base. Sumerian mathematicians used the number sixty, and computer scientists today use the number two.

With a base fixed, all numbers receive a standard description in terms of various powers and products of ten. Ten squared, or 10^2, is ten times itself. Ten cubed, or 10^3, is ten times ten times ten. Ten taken just once, or 10^1, is just ten itself, while 10^0 is one.

It is obvious—I hope—that any number can be expressed as a power and product of ten, so that seventy-three is $7 \times 10^1 + 3$, with even 7 and 3 subject to elimination in favor of 10^0 taken seven times to get 7, and then three times to get 3.

Seen in the light required for transit in one direction, the very first page of the dictionary contains and so reflects the essence of the scheme. Names in boldface are on the left; protected by the hedge of two parentheses, the numbers that they name are on the right:

0 NAMES (0×10^0)
1 (1×10^0)
2 (2×10^0)
3 (3×10^0)
4 (4×10^0)
.
.
.
9 (9×10^0)

The next few pages of the dictionary encompass the numbers between ten and nineteen. Addition has been allowed to supplement the work of multiplication.

10 $(1 \times 10^1 + 0 \times 10^0)$
11 $(1 \times 10^1 + 1 \times 10^0)$
12 $(1 \times 10^1 + 2 \times 10^0)$
.
.
.
19 $(1 \times 10^1 + 9 \times 10^0)$

The dictionary continues in this spirit, of course, page after page, and so name after name. It is for the most part the first dozen or so pages of the dictionary that are taught in childhood. No one really has need for exceptionally large numbers in daily life. Substantial private fortunes are easily encompassed by a few tens together with their exponents and heirs.

Although examples have been restricted to numerals of the form **ab,** the dictionary's scheme may be extended to encompass celebrity numbers requiring three symbols for their designation. Or more, as the case may be. Were the Devil inclined to **AEM,** he would assign **666** the following entry:

$$\mathbf{666} \; (6 \times 10^2 + 6 \times 10^1 + 6 \times 10^0),$$

remarking that positional notation has been extended to cover counting in terms of hundreds, as well as tens, and then ones: 6 (hundreds) 6 (tens) 6 (ones).

Give the Devil his due: He has gotten this right.

FROM THE BASE TEN

Transit now proceeds from numbers back to their names:

$0 \times 10^0 \; (\mathbf{0})$
$1 \times 10^0 \; (\mathbf{1})$
$2 \times 10^0 \; (\mathbf{2})$
$3 \times 10^0 \; (\mathbf{3})$
$4 \times 10^0 \; (\mathbf{4})$

.

.

.

$9 \times 10^0 \; (\mathbf{9}).$

If transit from names to numbers proceeds by the rule that the numbers correspond to powers and products of ten, transit in reverse is governed by the rule that the names of the numbers represent their coefficients. The numeral **666** is thus derived from $\underline{6} \times 10^2 + \underline{6} \times 10^1 + \underline{6} \times 10^0$, with underlining in this case calling attention to what is underlying, and that is the number 6.

The scheme works for larger and larger numbers; it is universal, the formula

$$a_n \cdot 10^n + a_{n-1} \cdot 10^{n-1} + \ldots + a_1 \cdot 10^1 + a \cdot 10^0$$

encompassing any number whatsoever, its name recoverable from its coefficients, $a_n \, a_{n-1} \, a_1 a$. The letter n in this formula serves as a bookkeeping tool, and so an index. It designates a number, one contingent on context. In the case of the number 666, n designates the number two, so that 666 is itself designated by a formula of the form $a_2 10^2 + a_1 10^1 + a 10^0$. There are three parts to the formula, but only twice is the underlying coefficient raised to a power of ten greater than zero. The number six is cited three times, the first to get to the hundreds, the second to the tens, and the third simply to get to six unadorned.

In all this, the elementary operations of **AEM** have been used to *explain* its notation, and its notation used to *express* its elementary operations. This is not a paradox: it is the way things are.

Like the human mind, the system of **AEM** cannot be completely dissected into its parts.

It stands as a whole.

12

In his introduction to Charles M. Doughty's Travels in Arabia Deserta, *T. E. Lawrence attempted to describe the character of the desert Arabs that both he and Doughty had admired. "They are the least morbid of peoples," Lawrence wrote, "who take the gift of life unquestioningly, as an axiom."*

RECURSION

It is a nice, a suggestive, way of putting things: to take life as an unquestioned gift is to take it as an axiom. A gift might be axiomlike because unquestioned; an axiom, giftlike because unearned.

What the stolid Bedouin might have made of this civilized juxtaposition, I do not know. The Bedouin seem to have regarded both Doughty and Lawrence as lunatics.

In his casual remark, Lawrence nonetheless touched a sensitive nerve. An axiomatic system expresses a sophisticated attachment to principle. It is not a game, nor should it be. The axioms of Euclidean geometry enter the mind as conjectures, but once entered, they compel allegiance in a way that no conjecture ever does.

Within an axiomatic system an initial gesture of assent generates a large and complicated system of allegiances, so that having accepted without demurral the Peano axioms, the mathematician finds himself later committed to the theorems that follow; he is often surprised by what his beliefs entail.

But what of the *definitions* of **AEM**? Are they mere verbal contrivances? If so, they invite the subversive thought that, no matter what they say, they could say something else.

Were the definitions of **AEM** theorems, all would be well. But the relationship between the axioms and the definitions of **AEM** is not an easy matter to pin down. The connection is elusive. The axioms of arithmetic say nothing about addition or multiplication; the definitions say everything, and no inference can connect nothing and everything.

How, then, can the axioms encompass the definitions?

And if they do not, what use are they inasmuch as it is the definitions that are doing so much of the work?

The definitions of addition and multiplication are definitions by descent. Computer programmers know the underlying technique as definition by recursion, as when a program allows a definable function to call its own values. That definitional descent is the common property of mathematicians and computer scientists is no reason, of course, to think it any good. Misery loves company.

But are there good reasons to think it bad?

That is another question entirely, and one that can be answered only by a more sophisticated sense of what might make these definitions good *or* bad.

Addition and multiplication take two numbers to a third; they have an active life, and *any* definition limning their nature must be justified in terms of what we know about the facts of life.

When it comes to particular cases, that justification is easily enough achieved. The definition of addition vindicates the proposition that the sum of four and three is seven. And in general, the definition provides its own witnesses. There are no surprises.

The fastidious intellect will wonder whether these witnesses might represent nothing more than the accidental intersection of a verbal artifact in the definition and a correct conclusion in its application.

The fact that definitional descent indicates that the sum of four and three is seven says nothing, after all, about the sum of *five* and three. That remains to be determined. Questions about how a definition by descent is to be extended are often answered by the useful phrase *and so on.*

But in order to justify *and so on,* mathematicians need more than definitional descent itself.

They need *proof* that definition by descent defines something in the world at large.

If the mathematician cannot bring the first into tight alignment with the second, what remains of the claim that he is the master and not the mastered?

CREDIT DUE

The American logician Stephen Kleene stated and proved the recursion theorem in a treatise entitled *Introduction to Metamathematics.* Though in some respects clumsy, and in many respects difficult, Kleene's book is nonetheless moving because it is the record of a man struggling with a great many daring new ideas. If Kleene lacked the temperament to make his thoughts elegant—perhaps this is because he was first concerned to make them disciplined.

By justifying definitional descent, the recursion theorem answers a need. A theorem of this sort—a theorem *in defense*—must do two things. It must demonstrate that definitional descent has some meaning in the real world; and it must demonstrate that whenever definitional descent is at work, the result is unique.

These demands are, of course, obvious. If there is nothing that definitional descent defines, the technique is useless. And if definitional descent defines more than one thing, *which* one regulates addition? Or any other operation?

A recursion theorem establishes once and for all that in definitional descent, there *is* a something answering to the definition, and that, moreover, it *is* unique. The question whether the definitions do justice to the facts is settled in the best possible way.

Justice is seen to be done.

DETACHMENT

Suppose that the number two has been encouraged by its cockpit crew to grow by exponentiation and to keep on growing. There is 2^0, 2^1, 2^2, and so on to 2^x, where x is any natural number.

Whatever the niceties of notation, the function 2^x demands a mental movement, an action, a taking of one number to another. This way of considering things is successful in reducing the idea of a function to a command: *take the number two and raise it to the power x.*

It is somewhat less successful in providing an analytically supple instrument by which the function might be described—*this* function, *any* function.

Although a function is an instrument of action, in acting it leaves behind a trace, a kind of characteristic signature. A certain intellectual detachment is required to see it, a contemplative willingness. The trace is evident in a function as simple as 2^x. Acting one at a time on the numbers, this function leaves its trace by means of the *pairs* of numbers it binds together: 0 and 1, 1 and 2, 2 and 4, 4 and 16, and so upward.

In this way, the function 2^x is *identified* with an infinite set of ordered pairs, with *the* set of ordered pairs, {<0,1>, <1,2>, <2,4>, ... }, in fact, the interior brackets indicating order, and the set-theoretic squiggle ({,}) binding the collection of ordered pairs into a single detachable object of thought. The set of ordered pairs is infinite, because when it comes to exponentiation, there is no end to the process, and so no end to the function.

In thinking of 2^x as an exponential call to action, it is the concourse between numbers that counts. The function 2^x raises two to the power x. *That is what it does.* With its role in action diminished, 2^x appears as a set of ordered pairs {<0, 1>, <1, 2>, <2,4>, ...}. *That is what it is.*

Questions about definitional descent now acquire a sharper focus:

Does definition by descent establish the *existence* of a function? And is it *unique*?

Inasmuch as the functions may now be identified with sets of ordered pairs, this is a very considerable improvement over the question whether definitional descent defines *something*.

THE RECURSION THEOREM

Like addition and multiplication before it, the function 2^x lends itself nicely to definitional descent.

Two clauses are required. The first forces the function to ground at zero:

1 $$2^0 = 1.$$

And the second provides for definitional descent itself by defining 2^x downward by doubling down:

2 $$2^{x+1} = 2(2^x).$$

Definitional descent works in this case as it has always worked, and that is to any finite extent.

Consider with perhaps a bit more care the line of code: $2^{x+1} = 2(2^x)$.

In a sense, what could be simpler? Or more evident? Two raised to the third power is *twice* two squared: $2^3 = 2(2^2)$.

What remains unnoticed is the curious way in which the expression $2(2^x)$ varies as x itself varies. The expression 2^x designates plainly a function, a concordance of variation, but if 2^x expresses a function, *then so does* $2(2^x)$. Just look at the variations *it* induces: if x equals 0, then 2^x equals 1, and $2(2^x)$ equals two. If x equals 1, then (2^x) equals 2, and $2(2^x)$ equals four. The function revealed by $2(2^x)$ is the faithful old function $2x$, here devoted, as it is everywhere devoted, to taking a number and *doubling* it.

The foreground having been grounded in 2^x, as art historians say, notice the equally important background, which consists of three parts: the natural numbers; the faithful old function $2x$, and the number 1.

The recursion theorem goes beyond the particulars of this landscape painting to encompass things in general.

The function (2^x) is thus replaced by the all-purpose $f(x)$. Beyond the fact that $f(x)$ takes numbers to numbers, its identity is unknown.

The function $2(2^x)$? It is demoted as well, and by the same demotion in particularity. Whence its replacement by the service-able and all-purpose function $g(x)$, a heavy lifter in the numbers-to-numbers trade, but a lifter indifferent to what he is lifting (*I double them numbers, I triple them. You want it I do it*).

Similarly, the number 1, now *any* number c.

Functional dependencies nonetheless remain nested, but at a higher level of abstraction, with g acting on f to form $g(f(x))$.

Having flown up into abstraction, the recursion theorem now comes down in assertion. *Whatever* the function g, the recursion theorem affirms, and for some number c, there *exists* a *unique* function f, one meeting the condition that $f(0) = c$, and that, moreover, $f(x + 1) = g(f(x))$.

It is in this way and by means of the recursion theorem that definitional descent achieves its apotheosis. *It works.* This is a good thing. Having taken my word for its importance, readers must take my word for its proof. And this, I suppose, is a bad thing. When a proof of the recursion *is* forthcoming, it is achieved at some cost. The theorem requires the resources of set theory. Functions must be treated by means of their identification with sets of ordered pairs. And it is complicated.

Even in **AEM,** what is elementary cannot always be justified by what is elementary.

. . .

WHAT IT DOES

The recursion theorem justifies definitional descent by drawing a connection between the recipe or algorithm embodied in definitional descent and the existence of a unique function, the one that definitional descent has presumably defined. A recipe is a form of words. The theorem serves to convey them to the world at large.

The recursion theorem is by no means absolute. Its conclusions are conditional. The function 2^x may be given something like absolute indemnity. It exists and it is unique. But the function $2x$ cannot be justified in the same way. Whether *it* exists and whether *it* is unique, the theorem does not say.

How could it? A theorem does what a theorem can, and no argument in mathematics manages to dispense with all its assumptions in order to relocate them to its conclusions.

Within **AEM**, of course, the function $2x$, since it expresses simple multiplication, may *also* be given a definition by descent, and the recursion theorem again brought to bear by virtue of still another function.

Where will it all end?

With the concept of succession, as it happens. That is the end of it—a fundamental idea, something that can neither be removed nor made *any* more tractable than it is.

If anyone were to ask how he might know that the operation of succession exists and is unique throughout the entire infinite range of the natural numbers, there is nothing to be said in response.

It is a matter of belief, an item of the faith.

13

At the beginning of the nineteenth century, all was not well at Cambridge or Oxford.

MEN OF THE LAW

One hundred years before, Isaac Newton had lent to the English mathematical community the immense force of his genius; in the years following his death, it seemed quite sufficient for English mathematicians, sniffling away the damp, cold winters, to say, when asked to justify their keep, that they were of the party of Isaac Newton. If oppressed by doubts, they assured one another that by devising examinations for undergraduates that were both difficult and pointless, they were keeping alive a sacred flame.

The German mathematician Carl Gustav Jacobi once visited Cambridge University. At dinner, he was asked who he thought was the greatest living English mathematician.

Arrayed around Jacobi were, of course, the mathematicians of the day, each man eager to see himself promoted to the High Table of international recognition, or willing to have his favorite promoted; they imagined telling their colleagues, *Why, at dinner Jacobi said that . . .* But nothing like that happened.

The greatest *English* mathematician? Pausing, I am sure, with the respectful sense that an honest answer could hardly please his hosts, Jacobi said simply, "There is none."

And as quickly sat down.

.   .   .

HOW THINGS CHANGE

In the early years of the twentieth century, the Scottish-American physicist Alexander Macfarlane delivered a series of lectures at Lehigh University in Pennsylvania. His purpose was hagiographic: he had come to praise. His lecture series was entitled *Ten British Mathematicians of the 19th Century.* Beginning with George Peacock, Macfarlane limned the nine other significant figures of the nineteenth-century British mathematical experience: Augustus De Morgan, William Rowan Hamilton, George Boole, Arthur Cayley, William Kingdom Clifford, Henry John Stanley Smith, James Joseph Sylvester, Thomas Penyngton Kirkman, and Isaac Todhunter. Of these, Hamilton, Cayley, Sylvester, and De Morgan were mathematicians of the very first order; Boole and Clifford very important; the others notable; but even the least among them capable. The drying up of English mathematics had ended within the century.

English mathematicians of the nineteenth century had all received a very similar education, one based on the study of classical languages. Except for Sylvester, who was an excitable hysteric forever fleeing some self-made shambles, they were controlled in their personalities. It is very easy to imagine them moving smoothly in political as well as mathematical circles, not because they wished to acquire power, but because they understood its use. They were worldly. And then again, almost all of the English mathematicians had some connection with the law. Arthur Cayley is today remembered both as a great mathematician *and* as a notable member of the English bar, a remarkably effective and discreet master of the various instruments of conveyance. A rich scene is thus suggested: A conveyance of some secret nature is at issue, one involving a considerable sum of money. Sitting before Arthur Cayley is a well-known public figure, an Eminence. He has entered chambers by a side door. With his demi-spectacles resting on the bridge of his large nose, Cayley gathers his papers together and, after tapping their bottom and then their side edges against his polished mahogany desktop, arranges them in a neat squared pile.

After rapping his walking stick just once in order to clarify Cayley's attention, the Eminence asks in a low, silky baritone: *Do we understand one another, Arthur?*

Allowing a small, delicate, deferential smile to play around his dry lips, Cayley murmurs: *Perfectly, sir.*

With his visitor departed by the same unmarked side door from which he had entered the room, and with millions of pounds now in play, Arthur Cayley, L.L.B., lays aside the instruments of conveyance with which he has been entrusted and, drawing his notes to the center of his table, resumes those very interesting calculations that he had been on the verge of completing before being interrupted by the demands of life.

AN UNATTACHED ENGLISHMAN

Augustus De Morgan was born in India in 1806. His father had been an official of the British East India Company, and his grandfather had also been born in India. With the outbreak of the Vellore Mutiny the very year his son was born, De Morgan *père* prudently repatriated his family to England. As an adult, De Morgan *fils* described himself as an unattached Englishman, a curious designation, one suggesting an incoherent division in the man's allegiances: *unattached*, but an *Englishman* nonetheless.

De Morgan's early education was erratic. His father having died when he was ten years old, De Morgan's adolescence was dominated by his mother's ardent devotion to the Church of England. She wished very much to see her son a clergyman, a career that in the early years of the nineteenth century would have secured him a modest income and a gratifying measure of respect. Exposure to theology seems to have left De Morgan indifferent to the religious vocation, perhaps because his disposition was ardent without being devout. He was given instruction in the classics by an Oxford tutor, and despite the now universal opinion that classical languages exhibit a grammar no more logical than that of modern languages, he acquired from his study a very balanced logical sense, one that served him very well throughout a career notable for its controversies as well as his research. De Morgan was not a great mathematician; but he was a pioneering logician.

THE FOURTH WRANGLER

At Trinity College in Cambridge, De Morgan came under the influence of George Peacock.

And no wonder. Many young men were attracted by Peacock's personality; they were attracted by his warmth. The man had the gift of suggesting to young men that everything was possible. Peacock was born to hustle, bustle, jostle, and command, but he had as well a clear-eyed sense of who in the English mathematical establishment could be counted on, who counted in, and who counted out. If he lacked mathematical talent, he disguised this very effectively by conceiving and then executing various reforms in mathematical education. It is a plan of action still popular today. Mathematical education has been undergoing reform since the time of the Greeks, and to precisely the same effect—which is to say, none whatsoever. It was Peacock who, together with Charles Babbage and John Herschel, formed the Analytical Society in 1815, and, by force of will and the strength of a hectoring personality, it was Peacock who managed to persuade the English mathematical community that it showed no disrespect to Newton's memory to give up Newton's notation for the calculus. If De Morgan was devoted to Peacock, he was nonetheless his own man; he admired at a distance, but he did not bend. Unlike the other men in Peacock's circle, he despised the reeking lavatories of college athletic life, and coolly declined to turn himself out for any sport. He played the flute and was by all accounts an exceptionally sensitive musician; he read as he pleased and studied as he felt. He found the cram courses needed to succeed in the Tripos examination in mathematics distasteful, and did not, I suppose, take them very seriously. He thus achieved only a fourth position among the Wranglers, men who had taken honors in the examination, a mark of distinction that, although not utterly ignominious, could hardly have been the source of intense satisfaction.

14

*The laws of arithmetic. It is a curious phrase, the more
so since the mathematicians who discovered them were
lawyers.*

PROCEDURAL MATHEMATICS

If the laws of arithmetic are laws in their name, they are procedural in their character. They control the flow; they are peremp-
tory: *you may* or *you may not*. They lack romance. With theorems
about the prime numbers, it is quite the other way around. Primes
may be divided only by themselves and the number one. The first
primes are two, three, five, seven, and eleven. There is no end to
them, as Euclid demonstrated, but their distribution is notably
irregular. Out where the numbers are huge and glowering, the
primes are sparse, almost as if they had no very great eagerness
to mingle with those bruisers. Greek mathematicians discovered
that the natural numbers could all be expressed in terms of the
powers and products of the positive prime numbers, a result that
is often designated the fundamental theorem of arithmetic. They
were fascinated by their discovery and talked endlessly about it.

The procedural laws, poor drabs, are otherwise. But they are
no less important. If they do not stir the blood, they regulate its
flow.

POWER PLAYERS

By long tradition, the laws of arithmetic encompass association,
commutation, distribution, trichotomy, and cancellation, making
five in all.

95

1

The associative and commutative laws both control certain reversals in the operations of **AEM** — *symmetries*, we may as well say, the word lurid enough to bring the physicists into open court, hoping to see something new.

We think ordinarily of symmetries in terms of the static and familiar plane of the human face. When divided down the nose, the human face folds out into two symmetrical halves, each a mirror of the other. The symmetries encompassed by the associative and commutative laws are otherwise; they are dynamic. They are symmetries of action, and so symmetries of time.

In circumnavigating a city block, it hardly matters whether one goes north, west, south, and then east, or west, north, east, and then south. If the trip is different, the result is the same, a matter of going somewhere but getting nowhere.

There is no need for *me* to inconvenience myself by trudging around the block.

The block in the form of a cardboard square can come to me, and at my leisure *I* can rotate *it* in space.

If block rotation offers an example of a perfect symmetry, it is easy enough to spot the places in the real world in which symmetry lapses because the order in which an activity is carried out matters, and often it matters a great deal.

A marine recruit who aims and *then* fires his M16 is doing two things: he is aiming and then firing. And as his drill instructor (DI) remarks, having just remembered his name, "You're on the ball, Monckton; you keep at it, maybe next time you hit the target." His companion at arms, disposed from sheer stupidity to fire *first* and *then* aim, is doing something else: *he* is firing and then aiming. If his DI is unwilling to offer him a few words of encouragement, it is because the accident has left him speechless. The activities of aiming and then firing, and firing and then aiming, although similar in their specifics, are radically unlike in their consequences.

It is this distinction that both the associative and the commutative laws express.

The associative law provides service to both addition and multiplication; it comes into play when three or more numbers are being added *or* multiplied, and it is designed to cancel the ambiguity that arises when even so simple a sum as 5 + 3 + 2 is calculated.

The ambiguity is this: The sum of 5 + 3 + 2 may be computed in two different ways. *First* by adding 3 to 2 and *then* adding the result to 5. Or, *second,* by adding 5 to 3 and *then* adding the result to 2. On *both* readings, the result is ten.

The operation of addition is indifferent to association.

You may add three numbers in two ways.

Go right ahead.

And, indeed, go a good deal further, if you feel like it, by allowing the associative law to encompass *all* of the natural numbers.

For any three of them, the associative law affirms, x, y, and z,

$$(x + y) + z = x + (y + z).$$

The associative law is forthright enough to comprise a guide to conduct. It clears away a latent ambiguity. But it is not obvious. It could have gone the other way, and in both division and subtraction, it *does* go the other way. Twelve divided by six and then by two is not twelve divided by three, as the associative law would suggest

2

The commutative law is like the associative law in governing role reversals within addition and multiplication. Five plus three and three plus five designate one and the same number. For any two numbers x and y

$$x + y = y + x,$$

the commutative law verifying once and for all facts of the five-plus-three-is-equal-to-three-plus-five variety. So long as addition and multiplication are at issue, the commutative law is not apt to raise any eyebrows, but, like the associative law, it lapses with respect to subtraction and division. A dynamic symmetry is again at work in the commutative law, for it affirms that two ways of doing something are the same. It makes no difference whether five steps are climbed first and three climbed second, or whether three steps are climbed first and five steps climbed second.

If the commutative law is so *like* the associative law, what is the difference, all things considered, between them? I am not sure that there is any very great difference at all. Both laws express a certain mathematical indifference to the order in which operations are carried out. The associative law expresses this indifference by the displacement of its parentheses; the commutative law by a reversal of the numerals used to express the law. A mathematician with a passion for consistent notation might well write the commutative law as $x + (y) = (x) + y$, thus allowing a pair of otherwise useless parentheses to carry out the role reversal that the commutative law expresses by a reversal of numerals.

3

The cancellation law conveys an air of triviality at odds with its important role in proceedings to come. That $2 + 3 = (1 + 1) + 3$ is not a proposition requiring much by way of soul-searching. The numbers $2 + 3$ and $(1 + 1) + 3$ are identical, and there is an end of it.

But, going further, the cancellation law says that if this is so then 2 must be equal to $(1 + 1)$. The cancellation law allows for common factors to be struck from various numerical identities, and so represents a primitive form of division.

Or, more generally, it says that, for any three numbers x, y, and z such that $x + z = y + z$, it follows that

$$x = y.$$

Multiplication follows suit.

Nonetheless, the cancellation law, although true of the positive numbers, is *false* as I have stated it. From the fact that five times zero equals eleven hundred times zero—both are zero, after all—it does not follow that five equals eleven hundred.

What holds for these particular numbers holds for all of them: $x0 = y0$ does *not* imply that $x = y$.

It is the number zero that acts as a snag when the cancellation law is drawn against the fabric of the natural numbers, evidence again of the curious role that this vapid little number plays within **AEM**.

4

The trichotomy law divides families into fighting factions, like a well-designed trust.

With respect to the numbers seven and five, *either* seven equals five, *or* seven equals five plus some other number, *or* five equals seven plus some other number. This is so obviously true that a law to this effect may seem supererogatory, and it was not until the end of the nineteenth century that mathematicians noticed the obvious, saw that it required proof, and provided the proof that was required.

What the trichotomy law enforces is the sense that this *is* so, and that it must be so for all the numbers. There are these three factions and there is no fourth.

For any two numbers x and y, it is either

$$x = y$$

or

$$x = y + u$$

for some number u, or it is

$$y = x + v$$

for some number v.

There is another way of putting the trichotomy law, one contingent on the concept of order.

For any two numbers x and y, either

$$x = y$$

or

$$x \text{ is greater than } y$$

or

$$y \text{ is greater than } x.$$

But the trichotomy law *fails* for multiplication.
Try it out and see.

5

The distributive law is the first law in which addition and multiplication are considered jointly and severally.

The expression $3 \times (2 + 5)$ calls for two operations to be done; and it divides the responsibility for doing them. Addition goes first, if only to give multiplication a number on which to work; and multiplication goes next, simply to get things over and done with.

What of distributing multiplication *over* the numbers two and five *before* they have been summed, so that $3 \times (2 + 5)$ is expressed as $3 \times 2 + 3 \times 5$?

In favor of distribution is the fact that $3 \times (2 + 5)$ equals 21, and so does $3 \times 2 + 3 \times 5$. As far as anyone can tell by looking at examples, this is the case throughout.

And that is what the distributive law asserts. For any three numbers x, y, and z,

$$x \cdot (y + z) = x \cdot y + x \cdot z.$$

The distributive law, do note, goes *from* multiplication *to* addition, but not the other way around. There is no distributing addition over multiplication: $3 + (5 \cdot 2)$ does *not* equal $(3 + 5) \cdot (3 + 2)$.

It is not even close.

15

Definitional descent raises the question how a verbal artifice encompasses an infinite operation. It is by means of the recursion theorem that the mathematician is able to say persuasively that all is well.

THE TRANSCENDENTAL FRATERNITY

But **AEM** goes well beyond the definitions of addition or multiplication. The associative law of addition ratifies common sense by confirming that $(3 + 5) + 2 = 3 + (5 + 2)$. No two numbers could be more sincerely equal than $(3 + 5) + 2$ and $3 + (5 + 2)$. The same exercise—*look and see*—might be conducted with any three numbers arranged associatively by an equation, and the result is always gratifying because it is always the same: the numbers *are* equal. Yet the associative law encompasses *all* of the natural numbers. Wherever one might look, and whatever one might see, some numbers must remain unlooked at, their associations unseen. Mathematicians have the power in the exercise of their art to go beyond the limit of what is finite. They are a transcendental fraternity. Their claims evokes anxiety and astonishment in equal measure:

Have they?
They have?
How?
And inevitably—
Show me.

· · ·

INDUCTION

Many of the proofs in **AEM** proceed by induction. So do many proofs in mathematics generally. The principle is easy to state and difficult to grasp.

Is some property true of the natural numbers? Is it true of *all* of them? Counting is of no use. There are too many numbers, too little time. The principle of induction provides an answer: *yes*, just in case the mathematician can take a trip of two steps.

The first step: establishing an inductive base by demonstrating that some property—the property of interest—is true of the number 1 or 0.

The second: establishing an inductive hypothesis by proving that *if* the property is true of an arbitrary number *then* it is true for the very next number.

Given a *demonstration* of the inductive base and a *proof* of the inductive hypothesis—given this, given these two steps—then the principle of induction sings out that the property is true of all the numbers. It is altogether true. Such is the transcendental fraternity's transcendental whoosh.

Some sense of the principle is very old. It enters into various of the propositions that Euclid demonstrated in the *Elements*, unacknowledged but influential. In the seventeenth century, Blaise Pascal assigned to mathematical induction the dignity of a name, calling it the method of finite descent. He did not feel obliged to do more. The great mathematicians who followed Pascal used mathematical induction as if it had all along been a companion at arms. In the end, it was Augustus De Morgan who grasped that mathematical induction was a principle of inference, and it was De Morgan who felt secure enough in the exuberance of his understanding to express the principle in essentially modern terms.

De Morgan *and* Peano, I should at once add, for the principle of mathematical induction is entirely encompassed by the fifth and last of the Peano axioms. The axiom has always had a divided nature, functioning both as a statement about what is and what is not the case, *and* as a rule or way of proceeding. It is in its second incarnation that Peano's fifth axiom justifies the use of mathematical induction.

FALLING DOMINOES

The standard illustration of mathematical induction is a row of dominoes. The row has a beginning in the first domino, but it has no end, the dominoes extended in space, one after another. The dominoes are upright. Then the first domino is toppled. What this image is intended to suggest—what it does suggest, that is the devilishly clever thing—is that, beginning with that first domino, a great wave goes through the uprights, so that, leisurely in their amplitude, the dominoes begin to fall, one after another.

Every now and then, the Internet contains videos of Chinese adolescents endeavoring to create ever greater domino trains, one of them, I seem to remember, extending for miles and set to topple from a lecture room at the University of Beijing to the very center of the Square of Heavenly Peace, where the very last falling domino was supposed to have struck a gong. Only an unhappy accident in which one of the judges—these horrible things are contests, after all—seems inadvertently to have tripped over a domino in the vicinity of the Bao Guo Temple and sent dominoes falling in *two* directions, prevented the group from achieving a world record.

The judge was, it is satisfying to recall, executed for wrecking.

Whatever his failure, the image remains, and with it the message it is intended to convey. *If* the first domino is toppled, and *if* toppling any domino *then* topples the next, *then* all of the dominoes must topple.

I have often wondered whether this claim is *physically* true.

Momentum is transferred in any domino train, and this might suggest that the dominoes would keep right on toppling; but momentum is always *lost* in any domino train as well, and this might suggest that, as the train is extended into outer space, the wave that it expresses would gradually slow, dominoes toppling until one, in the luxury of defiance, remains upright.

These suspicions express the inevitable limitations of a physical analogy to any mathematical operation.

THE RATCHET

The principle of mathematical induction is an axiom, and so it is an assumption. It may be derived from still further assumptions, but that serves only to exhibit a chain of logical dependencies.* In order to apply the principle, the mathematician must demonstrate that a base for inference exists, and he must prove an inductive hypothesis, but having done this, his conclusion that he has established something for *all* of the natural numbers remains what it always was, and that is an assumption.

If assumptions cannot be demonstrated, they can be otherwise assessed. Is the principle plausible? Does it carry conviction? Mathematicians, it is true, have long accustomed themselves to induction; they are untroubled by its audacity, but a sense that *this* axiom involves a form of mathematical prestidigitation remains common nonetheless. Hoping to hear that inferential *click* that signals an assertion beyond doubt, some readers may remark that so far as they are concerned, there is no *click* at all, the emotional security that should signal a sense of the self-evident nowhere to be felt.

Enter the Ratchet, a device intended to demonstrate the plau-

---

* The derivation follows. Keep reading.

sibility of induction without in any way attempting the impossible task of proving it.

The associative law of addition affirms that for any two numbers a and b and any number z

$$(a + b) + z = a + (b + z).$$

The Ratchet shows how the numbers may be ground into compliance with the associative law, one by one. It is an *illustration* of proof by induction, one especially vivid in revealing its hidden mechanism, steel teeth, discrete *ka-chunks*.

Ratcheting commences on the assumption that the mathematician has *already* provided an inductive base for the associative law:

$$(a + b) + 1 = a + (b + 1)$$

and that he has *already* demonstrated its inductive hypothesis:

$$If (a + b) + n = a + (b + n)$$
$$then (a + b) + n + 1 = a + (b + n + 1).$$

The symbols n and $n + 1$ designate any two consecutive numbers—seventy-four and seventy-five, for example.

So far, promise without performance. Teeth now commence to turn, the Ratchet showing that association is true for the number *two*.

Those teeth in action: association is conditionally true for any pair of consecutive numbers. This is the burden of the inductive hypothesis.

So it is certainly true of the numbers one *and* two.

But association *is* true for the number one by the assumption of an inductive base.

So it *is* true for the number two. It is true *logically*.

Here is the inductive hypothesis coordinating one and two:

$$If \ (a + b) + 1 = a + (b + 1)$$
$$then \ (a + b) + 2 = a + (b + 2).$$

And its inductive base:

$$(a + b) + 1 = a + (b + 1).$$

The turn of the screw:

$$(a + b) + 2 = a + (b + 2).$$

Pure logic is at work, a principle known as Modus Ponens: from P and if P then Q, it follows that Q.

If association is true for the number two, then by just the same reasoning, it must be true for the number three. If true for three, then four, and if four, then five.

Now fully engaged, the Ratchet chews its way up the chain of natural numbers, one step at a time, an indefatigable logical machine, modest in its mechanism, impressive in its achievements.

Is the Ratchet a proof of association? No, of course not. The proof of association comes *before* ratcheting.

A susurrus of discontent.

Uh, it doesn't really show anything, is that what you're saying, Dr. Berlinski?

Half right: it doesn't really *prove* anything.

But right in a deeper sense, too, for the Ratchet is an instrument of revelation as well as an illustration. When it comes to the principle of induction, the mathematician's allegiances are to his transcendental fraternity. A proof by mathematical induction is a leap, one curiously akin to a *leap of faith*. It is a *leap*, because induction involves a crossing over from what is finite to what is infinite. It is a leap of *faith*, because as the Ratchet reveals, induction *never* goes beyond what is finite.

A *proof* of association displays the mathematician's helmet of gold, the Ratchet, his feet of clay.

Leaps of faith are what they are. There is no way to make them better or more secure.

And this, too, is important.

WELL ORDERED

Peano's fifth and final axiom divided its nature between a description and a rule. It was the rule that emerged as the mathematician's friend—often, indeed, his best friend—because, expressed as the principle of induction, it afforded the mathematician a way of making and then demonstrating claims about all of the numbers. Proofs are his business, and without the means of making them, where would he be?

The plausibility of induction as a technique of proof depends, as it must, on the assertion that is embedded in Peano's fifth axiom: *if* a set contains the number one, and whenever it contains a given number it contains its successor, *then* it contains them all.

It is by no means obvious, this assertion, and this is one reason that it was not until the nineteenth century that it was properly formulated. Once formulated, it *seems* obvious, because deep down it expresses the purely human conviction that succession, if allowed to continue, would generate all of the natural numbers.

The principle of mathematical induction, curiously enough, may itself be derived from a still more general assumption, an axiom in one system appearing as a theorem in another.

The setting is set theory, and the general asumption is the well-ordering principle. A set is well ordered if every one of its non-empty subsets contains a smallest element.

The natural numbers are an example—a perfect example in view of their importance. The even numbers, 2, 4, 6, 8, . . . , contain in two a smallest element. The odd numbers contain in one a smallest element. Even numbers and odd numbers are subsets of the natural numbers, parts to wholes, pieces to pies. What is more, the natural numbers as a whole contain either in zero or in one a smallest element. So they are well ordered.

But the negative numbers are *not* well ordered, confirmation, if any were needed, that they are in some obscure sense not well at all. The negative numbers begin at -1, but for any number n, they always have, in $n - 1$, a negative number smaller than n.

From the well-ordering principle, Peano's fifth axiom follows as a theorem, and it follows almost at once.

Peano may be allowed to repeat himself for one last time: a set S of positive integers that includes 1, and that includes $n + 1$ whenever it includes n, includes all of the positive integers.

The proof that this is so, *given the well-ordering principle*, is nicely conceived because it establishes the point at issue by means of a contradiction.

Is the principle of induction *false*, so that S does not contain all of the positive integers?

Very well. Consider the set K, comprising the positive numbers *not* in S, the *leftovers*.

If there are no leftovers, the proof is over.

Suppose, then, there *are* leftovers. In that case, by well-ordering, K has a smallest element in m.

Is that number 1?

Impossible. By assumption, one is over there, in the *other* set, S.

But, then, m must be greater than 1.

And at the same time, $m - 1$ must be a positive number. There is, after all, *no* positive number between zero and one.

But look, that means that $m - 1$ is in S too.

So? It is in S.

But if $m - 1$ is in S, then so is m, since m is simply $(m - 1) + 1$. This is enough to establish a quick contradiction.

Going left: the number m is in K.

Going right: the number m is in S.

One and the same number cannot be in both S and K.

Does this argument do very much to establish the principle of mathematical induction? Or does it represent a contrivance?

The latter, I suspect.

The principle of mathematical induction talks directly to the natural numbers: it is about them, and about nothing else. The well-ordering principle carries the cold assurances of an alien world in which sets are the chief thing. But it goes without saying, I hope, that the mathematician in need is perfectly prepared to overlook aliens in his crowd and consider the well-ordering principle a decent fellow after all.

16

In the midst of these dry details, it does not hurt to be reminded of the passion that they conceal, and the drama they evoke.

ARDOR

Sofya Vasilyevna Kovalevskaya (Sonya Kovalevsky) was born in Moscow in 1850 and died in Stockholm in 1891. The city is mathematically unlucky, René Descartes having died there in 1650 of some dreadful bronchial infection. A woman of very considerable talents, both as a mathematician *and* as a writer, Sonya Kovalevsky lived within the confines of an impudent Russian melodrama, simultaneously its heroine and its victim.

Within that melodrama, there was wealth, privilege, and a luxurious estate; there was an overbearing father, a man whose moods could ruin the household's peace; there was a most musical *mama*, the daughter of a famous Russian astronomer; there was an older sister, Anya, firstborn and so best loved, and a younger brother, Fedya, the household prince and heir, Big Anya and Little Fedya attracting dangerously unstable emotional affinities from their parents; there was a strict, prim, and humorless governess, mad for decorum and discipline—*of course* there was; and there was that staple of every Russian melodrama, an eccentric but fun-loving uncle, who told an eager, unloved child fairy tales, arranged a chessboard to suit her pudgy fingers, and talked with great dreaminess about "squaring the circle, asymptotes, and other things that were unintelligible to me and yet seemed mysterious and at the same time deeply attractive."

As a child, Sonya Kovalevsky acquired the rudiments of

nineteenth-century mathematics by studying a textbook written by yet another Russian figure scuttling in from the theater wings of time, a Professor Tyrtov, who just happened to be a landowner, a man of means, and a neighbor, his conviction that women were incapable of mastering mathematics dissolving in helpless admiration as the shy but determined Sonya Kovalevsky deftly sorted through the complicated formulas of his textbook and solved the problems that it presented. Having discovered her talent, Tyrtov persuaded her father that she must be allowed to continue her education; Sonya Kovalevsky becoming their communal ward, a little innocent being transferred from the care of one well-meaning, wise guardian to another.

For all that, her father's assent required four years before it was fully forthcoming, but in the end, and with the sense that he had done a manful but difficult thing, he allowed Sonya Kovalevsky to study analytic geometry and the calculus in Saint Petersburg. She was tutored, of course, and chaperoned, and kept cozy, comfortable, and captive; the distance in her life from the ordinary world in which men freely took up their studies in large, noisy, boisterous groups serving only to inflame the intensity of her desires, her pitiful pained ardor.

No one doubted that Sonya Kovalevsky was remarkable — not her Russian tutors, at any rate. And no one doubted that she deserved a university education. But Russian universities were closed to women. If Sonya Kovalevsky could not study at home, she would have to study abroad. In nineteenth-century Russia, as in contemporary Islam, an unmarried woman's freedom to travel was almost as difficult to obtain as her freedom to study, if only because her latent erotic power was considered so dangerously unstable a force that any father would be made uneasy by the thought of *his* darling daughter reclining with easy indolence against the cushions of the international sleeper departing from Saint Petersburg every evening, her decorously shielded limbs a provocation to plump Russian businessmen, military officers, card sharps, landowners, bureaucrats, Swiss officials, and even the ministers of various tea and pastry wagons.

A woman sitting alone and—of all things!—reading a treatise on mathematics was widely regarded among even educated men as an invitation to debauchery. Anna Karenina had spent a good deal of time traveling alone on the night sleeper from Saint Petersburg to Moscow, after all, and even though she was a married woman, no one could miss the alliterative clack of trains, traveling, time, and treachery.

What Sonya Kovalevsky might have done abroad while living alone and doing as she pleased was an exercise in her family's already agitated erotic imagination. The solution was a masterpiece of contrivance: an arranged marriage to one Vladimir Kovalevsky, a biologist by training, a paleontologist in prospect, and an ardent admirer of Charles Darwin. By surrendering her liberty, Sonya Kovalevsky gained her freedom. She decamped for Heidelberg, a beautiful university town in the nineteenth century, and one blessedly untouched in the twentieth.

Her professors' glowing testimonials enabled her to meet Karl Weierstrass, one of the eminences of the German mathematical academy. Kindly, rumpled, and disheveled, Weierstrass challenged Sonya Kovalevsky with a set of problems he had prepared for his advanced students, and when she had solved them with a positively alarming degree of ease, he determined generously that her "personality was [strong enough] to offer the necessary guarantees" for advanced training. To the uncles she had already acquired, Sonya Kovalevsky added a powerful new uncle, so that she appeared in European mathematical circles as the glowing star at the center of an avuncular galaxy.

Thereafter, her short life was consumed by her ardent nature. The contrived and pathetic marriage into which she had entered as a matter of convenience made demands of its own, and both she and Vladimir Kovalevsky discovered to their surprise that an arrangement to which neither was committed became one in which both were consumed.

After four years in Heidelberg, the couple returned to Saint Petersburg, where Sonya Kovalevsky discovered almost at once that a society unwilling to allow her an education was equally unwilling to afford her a position. She gave birth to a daughter, whom she seemed equally to have adored and to have neglected. She wrote for various theatrical and literary publications; she started a novel. Persuaded, like so many other talented women, that her gifts were fungible, she and her husband embarked on a number of business schemes, each one a notable, even a spectacular, failure, disasters accumulating until their marriage dissolved under the strain.

Vladimir Kovalevsky took his own life in 1883.

It can hardly be said that Sonya Kovalevsky lived a life without honors—only that she lived it without luck. Decamping from Saint Petersburg for Paris, where her sister was already making the acquaintance of various revolutionary bohemians, men whose commitment to violence was offset by their indifference to work, she re-entered the mathematical scene, and, with that special gift she had for attracting uncles, caught the eye of Gösta Mittag-Leffler, a student of the great Weierstrass, and a powerful and determined mathematician in his own right. Mittag-Leffler became her last champion, in the end persuading the University of Stockholm to award her a probationary position, one of those awkward arrangements so familiar in academic life in which every requirement except decency is satisfied.

She continued to work; she achieved many notable results in the theory of ordinary and partial differential equations, and in 1888, she received the *Prix Bordin* from the French Academy of Science. Like the Russians, the French were prepared to honor achievement without ever making it possible. Her position in Stockholm was made permanent; and she was elected to the

Russian Academy of Science. Her hope that as a member of the academy she might be rewarded by an academic position was not fulfilled, circumstances that she met with a characteristic mixture of contempt and resignation. In 1891, she died quite suddenly after suffering from pneumonia, and now survives as a face engraved on a Russian postage stamp, and a name attached to a crater on the far side of the moon.

All this belongs, I suppose, to the universal history of sadness; but in her autobiography, *A Russian Childhood*, Sonya Kovalevsky recalls with some sense of wonder an early memory.

She was eleven. Her bedroom required wallpaper, and for reasons that even Sonya Kovalevsky cannot explain, the walls were covered with notes and scribbles from a tour of the calculus owned by her military-minded father. Her uncle had already introduced her to mathematics, but not to higher mathematics or the formulas of the calculus.

"I noticed certain things," she wrote, "that I had already heard mentioned by uncle. It amused me to examine these sheets of hieroglyphics whose meaning escaped me completely but which, I felt, must signify something very wise and interesting."

But, really, isn't this how we all are, much impressed by things we do not understand and hoping that they represent something very wise and interesting?

17

The associative law of addition says that for every number
z, and any two particular numbers a and b, a + (b + z) =
(a + b) + z.
If this is what it says, here is how it is proved.

THE PROOF

The proof of the associative law of addition proceeds mathematically by induction.It is thus necessary to show that the law is true for one (or zero) and then necessary to demonstrate an inductive hypothesis. The proof depends on a number of bookkeeping identities, small matters of detail, not at all swashbuckling.

A word first about notation. The symbols a and b are often called parameters, and this in order to distinguish them from variables, such as x and y. Unlike the variables, which are variable, the parameters are intended to designate particular numbers. These symbols afford the mathematician an enhanced sense of specificity, as when an attorney, constructing a hypothetical case for the jury, says, *Suppose A walks into a bar and gets into a fight with B.* No one hearing him out is apt to wonder just who A and B might be, but there is nonetheless some presumption in favor of A and B as individuals. The variables x, y, and z, logicians say, range over all the numbers; the parameters a, b, and c, they add, designate some number or other. Sometimes parameters are used to hold some part of a mathematical expression tied to specific numbers, while the variables allow other parts to roam. The expression ax is like that, designating *any* number x and *some* number in a.

These are distinctions of convenience. They are not of the

essence. The proof of the associative law of addition would proceed just as well were it written $x + (y + z) = (x + y) + z$. But the parameters have an advantage denied the variables. Somehow they seem cheerier.

The proof then: there is, first, Peano's fifth axiom that any set of numbers that contains the number 0, and that contains the successor of any number *if* it contains the number itself, contains all the natural numbers.

It is thus entirely reasonable to suppose that there does exist a set of numbers A that satisfy the law of association.

What we do not *yet* know is whether *this* set, *our* set, *the* set A, contains *all* of the natural numbers.

Plainly, the number zero belongs to A. Whatever the numbers a and b, the numbers $a + (b + 0)$ and $(a + b) + 0$ are one and the same. "The fact [is] notorious and indubitable," as Dr. Johnson remarked in another context, "so easy to be proved that no proof [is] desired."

This observation provides the inductive base of an inductive argument.

The inductive hypothesis remains; and it must be demonstrated:

If z belongs to A, then z + 1 belongs to A as well.

This is the fulcrum of the argument, the place where leverage is gained after force is applied.

In order to establish the inductive hypothesis, it is necessary to assume that z itself belongs to A, and this for *any* number z.

The assumption that the number z belongs to A is made *for the sake of argument*. At issue, after all, is a hypothetical proposition: *if* a number z belongs to A, *then* $z + 1$ belongs to A as well. To demonstrate a hypothetical proposition, it suffices to assume its antecedent—z belongs to A—and then derive its consequent:

$z + 1$ belongs to A as well. The antecedent is assumed, as logicians say, *conditionally*, in order to show that the hypothetical is *as a whole* true.

Since the number z belongs to A by assumption, it follows that

$$a + (b + z) = (a + b) + z.$$

What holds for the number z must hold for the very next number:

$$a + (b + (z + 1)) = (a + b) + z + 1.$$

Prove it demands the voice of common sense.

Proceed booms out the law of logic.

Very well. The jury is reminded that what is at issue is an *identity*, one asserting that $a + (b + (z + 1))$ *equals* $(a + b) + z + 1$.

And reminded again that in order to establish that two things are equal, it suffices to begin with one of them and then derive the other by a chain of identities.

So begin with

$$a + (b + (z + 1)).$$

And notice that $a + (b + (z + 1))$ equals

$$a + ((b + z) + 1)$$

by the definition of addition.

But $a + ((b + z) + 1)$ equals

$$((a + b) + z) + 1.$$

because z is a member of A.

But then $((a + b) + z) + 1$ equals

$$(a + b) + z + 1,$$

again by the definition of addition.

So by connecting identities in a logical daisy chain, it follows that

$$a + (b + (z + 1)) = (a + b) + z + 1,$$

All done.

And so quickly too.

This proof may well be difficult to grasp, but it is by no means difficult. Difficulties reflect nothing more than a peevish refusal on the part of short-term memory to retain the details long enough for the point of the proof to emerge.

Let me just refresh everyone's short-term memory. The point of the proof is to show that association is true for all of the natural numbers. That much everyone remembers.

And everyone remembers that the proof proceeds by induction: True for zero first—show that. And then show that association holds for the next number up, $(n + 1)$, *if* it holds for the last number down (n).

The techniques by which the proof proceeds? Conditional assumption first: Suppose that association is true for the number n. And a daisy chain of identities next, one connecting the assumption that association is true for n with the conclusion that it is true for $n + 1$.

The largest gestures of the proof are thus: state the thing, assume the antecedent, remember the definitions, come to the conclusion.

What could be simpler?

But having said that, let me at once add this. What could be more complicated?

In order to prove the associative law of addition, an appeal to the definition of addition is required. Very well. The appeal has been made. But in order to justify the definition of addition, the thoughtful observer must keep the recursion theorem in mind. Very well. It has been kept in mind. But then, in order to go further, the proof must proceed by induction. Very well. Proceed by induction. But in order to proceed by induction, we need Peano's fifth axiom.

And, no, there is in this nothing unacceptable, but what is unexceptionable is very often instructive, and that is the striking, even remarkable, contrast between the plain obviousness of the associative law of addition, and the delicate and detailed structure needed to bring it squarely into the tent of things that are so because they have been demonstrated.

Lawyers and logicians have been talking about simple addition, after all, and saying something of no more depth than that $2 + (3 + 5) = (2 + 3) + 5$. No one has ever doubted that this was so.

The mathematician alone cannot leave well enough alone.

And now neither can you.

18

The natural numbers are what they are; the number zero is what it is. They have a shadowy existence, one without shape.

THE OTHER SIDE OF ZERO

Nonetheless, they have a very natural geometrical representation, a *depiction*. A point is given in the plane, and from the point, a straight line extended, what geometers call a half-line. The point corresponds to zero. A unit is next marked on the line, an initial interval corresponding to the number one. The question of how big the unit does not arise. In what further unit could it be measured? With a unit given, the rest of the line is subdivided by multiples of the unit, the second, third, and fourth corresponding to the numbers two, three, and four.

This is the geometric way of thought: until we have *seen* something, we do not know it for what it is. Point and line bring the numbers to life by bringing them to ground. The idea is old. *We are here*, a Sumerian general might have grunted five thousand years ago, stabbing at the ground with a stubby forefinger, *and we go there*, the forefinger scooting aggressively in a straight line through the desert sand. A half-line conveys dramatically what the numbers themselves only suggest, and that is the promise of a new beginning. Wars, babies, drug-recovery programs, civilizations, and the universe itself are measured from a zero point; revolutions often change traditional calendars to encompass a year zero, when first the great things were undertaken. In addition to its other offices, the number zero, popping up insouciantly on an otherwise empty plane, suggests the ultimate of metaphysical mysteries, the way in which something appears from nothing.

Only a brief step separates geometry from cartography, the number zero appearing on maps as the nub or hub from which trips begin. The resulting half-line corresponds to the highway that, when imagined, stretches effortlessly into the blue distance; and when realized in real life, of course, it is inevitably subdivided by a succession of motels where at night trucks or toilets may be heard downshifting, often in alternation.

If the half-line embodies the desire to register things from their beginning, it almost always suggests a mental movement sunny in its disposition. The *other* side of zero conveys to this day a shivery suggestion of descent into darkness, the more so when the number line is stood on end. Going up from Paris's Charles de Gaulle Airport, we ascend through the ambient atmosphere and we ascend toward the light, the great lumbering Boeing 747 breaking free of the clouds to soar at last, but in crossing zero for the dark side, we descend by means of claustrophobic mine shafts, rat holes, or badger burrows, all of them going downward until they reach the bowels of the earth.

No one ever *descends* toward the light.

THE DARK SIDE

The negative numbers are the numbers $-1, -2, -3, \ldots$ They have always provoked unease.

The equation $4x + 20 = 0$ says that four times some number plus twenty equals zero. *Which* number might that be? The question hardly seems the occasion for mystification. A simple identity is at work in $4x + 20 = 0$, one antecedently familiar from the very similar equation $4x - 20 = 0$. But the equation $4x - 20 = 0$ has an obvious solution in the number five. Four times five minus twenty is zero. Whatever the solution to $4x + 20 = 0$, it is not obvious.

The Greek mathematician Diophantus, like Euclid a man of Alexandria, considered $4x + 20 = 0$; he could see at once that, *if* the equation had a solution, the *only* solution that it had was negative. His divided inner voices can no longer be heard, but they can be imagined:

Let me see . . .

Subtract twenty from both sides of the equation.

Because?

Because it cannot hurt.

Can I do that?

Why not?

I'm just asking.

Equals added to equals are equal. Or maybe it's the other way around. Anyway, it's in Euclid.

That leaves $4x = -20$.

*What next? I mean besides giving up. That **always** works.*

Maybe divide both sides of the equation by 4. That's supposed to work too.

And you get what?

Well, the deuce of it is, I don't seem to get anything but -5.

That can't be right, right?

On the other hand, it can't be wrong either.

Why not? Most things are. . . .

In the face of these difficulties, Diophantus did what so many mathematicians (and students) were afterward to do: he rejected the negative numbers as absurd, but with the uneasy sense that mathematics might, after all, require absurdities.

He was hardly alone. Brahmagupta used the negative numbers correctly in the seventh century, but always with the sense that the things were unclean and would, when insight was finally vouchsafed, be eliminated altogether.

Five centuries later the great Bhaskara had grasped how to find the negative roots (or solutions) of certain quadratic equations,

such as $x^2 = 25$, but, with the distinct archness of tone characteristic of a man who feels compelled to bow to public opinion, he
rejected his own conclusions, because "people do not approve of
negative roots."

Are these examples anachronistic? Not at all. In the late eighteenth century, one hundred years after the creation of the calculus, Francis Maseres, an English mathematician, wrote that
negative numbers "darken the very whole doctrines of the equations and make dark of the things which are in their nature excessively obvious and simple."

BAD LUCK, PIERRE

The negative numbers are today taught to children. It is said that
children take them in their stride (hardly a great recommenda-
tion, one might think). They may, those numbers, be negative,
but they are no longer dark, especially when the full number line
is allowed to supplant the half-line. Zero and the natural numbers are where they were, but the other side of zero opens to the
highway of another half-line. It is on this half-line that the negative numbers are inscribed. French highways now place crosses
commemorating fatal accidents where on the line the negative
numbers would go.

That **was** bad luck, Pierre. And in your new Porsche too.

An attractive symmetry is at work, in both the picture and its
interpretation. If the line were folded at its origin, positive and
negative numbers would undertake annihilation and creation, as
in quantum field theory. "Think of one and minus one," John
Updike once wrote. "Together they add up to zero, nothing, *nada,
niente*, right?"

Annihilation.

"Picture them together, then picture them separating, peeling
apart. Now you have something, you have two somethings, where
you once had nothing."

Creation.

Whatever the mysteries of creation and annihilation, the interpenetration of number and line suggests an improving relationship between arithmetic and geometry. Questions might be asked about how the numbers find their places on the line, the process of *fixation*, to use a darkroom term, but these questions are less important than the interpenetration itself, the way that arithmetic informs geometry and geometry arithmetic.

This might suggest a form of unity between arithmetic and geometry, one lying far beyond the resources of **AEM**, distinctly modern in the way it undermines established categories, but old in the way that it hints at the Indivisible, the Ineffable, the One.

DISTANCES

Starting at zero, a knight on horseback takes off on the number line, covering ground in the direction of numbers that are positive. He is, with loud aggressive clip-clops, following the sinking sun.

Let me draw the poor fat fool back to zero and point his horse in the opposite direction. He is now clip-clopping toward the rising moon.

It hardly makes a difference which way he goes, does it?

Zero to begin with; start out, strike out, and cover ground, riding toward the sinking sun *or* the rising moon:

> *Reiten, reiten, reiten,*
> *durch den Tag, durch die Nacht, durch den Tag.*
> *Reiten, reiten, reiten.**

---

* "Riding, riding, riding, through the day, through the night, through the day. Riding, riding, riding." The poem continues: "*Und der Mut ist so müde geworden und die Sehnsucht so gross*" (And courage has grown so tired and the longing looms large). (Rainer Maria Rilke, *Die Weise von Liebe und Tod des Cornets Christoph Rilke.*)

By the time he has developed hemorrhoids, Rilke's knight has covered one hundred miles whatever the direction he has ridden. How *far* a man on horseback goes is a matter of how far he has gone, and not which direction he has taken.

But consider now another rider, a mathematician. Let me just ease this lummox into the saddle. Riding toward the setting sun, he covers one hundred miles; but, returned to zero, and now riding toward the rising moon, the mathematician, unlike other men, covers *minus* one hundred miles. Minus one hundred is less than one hundred. It is less than one hundred by two hundred miles.

Whatever that mathematician may think, *he* is not riding along the number line.

Distances, like probabilities, cannot be negative.

DEBTS

Debts have long been the place where negative numbers are thought to be indispensable. Bankers with beady eyes were often heard to say that this was so, often in the movies. Real bankers, we now know, have of late obliterated the negative numbers from their deliberations, on the grounds that these might remind them of their debts, an odd way of confirming the importance of the negative numbers. Merchants throughout the ancient and medieval world knew perfectly well, of course, how to keep track of money coming in and money going out, but the association between debts, considered as a fact of life, and numbers, considered as an aspect of mind, did not emerge with any great clarity until the publication of Luca Pacioli's *Summa de Arithmetica, Geometria, Proportioni et Proportionalità (Treatise on Arithmetic, Geometry, Proportions and Proportionality)* in 1494. Pacioli ranged widely over fifteenth-century mathematics, but he was a practical man as well as a mathematician, educated by merchants and a tutor to their sons, and his treatise is still well known for its description of double-entry bookkeeping, the so-called Venetian method. As the name might suggest, double-entry bookkeeping demands that accounting transactions be recorded twice, once to mark the ledger from which the money comes, a second time to mark the ledger to which it goes. *

Do not go to sleep, Pacioli warned, *without making sure that debits and credits sum to zero.*

This they can only do if those debits are *negative*.

Negative numbers are useful in double-entry bookkeeping systems. And useful generally. My ledgers *and* my life display a good many negative numbers, often followed by exclamation points. But what is wanted from these examples is not so much proof that negative numbers have a use as evidence that they enjoy an identity. As far as bookkeepers are concerned, a debt of five dollars, or −$5, could equally well be recorded as D$5, with no expectation that new numbers have been in this way created.

Why *would* I require a new number to describe an old debt? The same number used when five dollars are owed me does perfectly respectable work when I owe them. If I have already spent those five dollars, then I have five dollars less than I had, and if I had nothing to begin with, then what I spent was not mine, but in any case, *it* was spent, and *what* was spent must be designated by the same natural number on my side of the ledger as might appear on my creditors' side. If the schnook from whom I borrowed five dollars has one number in mind in asking for his money back, and I have another, what is *it* that I owe *him*?

Nothing, I hope.

This is the paradox, then, of the negative numbers. To the extent that numbers are a measure of quantity, and so an answer to the question *how many*, they cannot be negative.

And if the negative numbers are not a measure of quantity, why think of them as numbers?

Was it Francis Maseres who wrote that negative numbers "darken the very whole doctrines of the equations and make dark of the things which are in their nature excessively obvious and simple"?

I believe it was.

19

Addition adds numbers to numbers; subtraction takes numbers from numbers. Taking away undoes adding up.

TAKE AWAY

A system in which things add up but do not add down is unbalanced, and reflects more the contingencies of a physical world in which hardly anything ever adds up than the niceties of **AEM**.

Like addition, subtraction is an operation, one working to measure the difference between numbers x and y instead of their sum. That difference mathematicians (and everyone else) write as $x - y$, the symbol $-$ formed by taking away one arm from $+$, that dwarf left with his outstretched arms but no feet or head, the symbols recapitulating precisely the operations they were meant to designate. Subtraction takes something away. The traditional notation for subtraction is nonetheless unfortunate, because the sign for subtraction and the sign for the negative numbers is the same. The symbol -5 designates the number minus five. The symbol $-$ works one on one. It changes the sign of the number five. But when the very same symbol appears between two numerals, as in $10 - 5$, it conducts a two-man operation, one indicating that two numbers are bound to a third. When symbols are combined so as to say that minus five is being subtracted from plus ten, the result is $10 - -5$, an assertion on the face of it that could be a conveyance in Morse code as well as an imperative in mathematics.

And what of $-10 - -5$, which suggests nothing so much as someone gagging in a steak house.

For heaven's sake, Doris, stop mumbling.

Ambiguities of this sort represent blemishes; they are often vexations; but they do not run deep. With the proper use of parentheses, merchants and mathematicians find it easy to explain what they mean, so that $10 - -5$ reappears when cleaned up as $10 - (-5)$.

But no sooner has $10 - -5$ been tidied to $10 - (-5)$ than $10 - (-5)$ reappears as $10 + 5$, thus suggesting that, in addition to its other duties, the two-man operation of subtraction, in which two numbers go to a third, has a sideline in one-man jobs and is busy changing a negative to a positive sign.

It is quite true, of course, that the number $10 - (-5)$ *is* equal to the number $10 + 5$.

But not because the sign for subtraction is busy doubling as the sign for negation. *It* has its hands full.

The traditional symbolism remains what it always was. It is clumsy.

A BROKEN SYMMETRY

In adding six to ten, the direction is up, and it is up in steps of one: ten, eleven, twelve, thirteen, fourteen, fifteen, at last, sixteen.

Six plus ten is sixteen.

The number sixteen having been given, subtraction works to peel away the same six numbers. The direction is down: sixteen, fifteen, fourteen, thirteen, twelve, eleven, ten.

Sixteen minus six is ten.

Even in ordinary English, these arithmetical trivialities suggest a form of symmetry at work—

Six plus ten is sixteen.

Sixteen minus six is ten—

one that goes backward as well as forward, with the change prompted only by the substitution of **minus** for **plus**, and the switch of number nouns in their position.

Subtraction has in all this remained an intuitive operation, one not yet defined. But the definition is easy enough. The difference between two numbers is a number in turn. Like addition, subtraction is an operation that takes two numbers to a third. The difference between sixteen and six is ten.

For *any* two numbers x and y, their difference $x - y$ is a number z such that $z + y = x$. From the definition, particular cases follow. Let x be sixteen and y six. Their difference in z is the number ten, and ten, moreover, plus six *is* sixteen.

Only addition is at work. The subtleties of subtraction as an independent operator in **AEM** lie encoded in alien seed.

After all, that sixteen *minus* six equals ten *means* that six *plus* ten equals sixteen. If it does not mean this, it does not mean anything. What could better reveal the symmetry between addition and subtraction than the discovery that subtraction may be subtracted from **AEM**, leaving only addition behind?

THE SYSTEM OF THE INTEGERS

The negative numbers begin at -1, and, like the positive numbers, they go on forever.

Edmund Landau introduced the negative numbers by saying: "We . . . *create* [my italics] numbers which are distinct from zero as well as the positive numbers. . . ."

We *create*? Uh-oh.

The creation to which Landau is committed is as unavoidable as it is high-handed. Without the negative numbers, there is no subtraction, and without subtraction, no symmetry either.

It is the negative numbers that offer subtraction full parity with addition. They restore addition to itself. The operation it commands makes sense coming up or going down. For *every* x and y, the operation $x - y$ designates the one and only number z such that $y + z = x$. If before six minus sixteen yielded nothing, now it yields something. It yields the number that when added to sixteen is six. That number is minus ten. No other number will do. Six minus sixteen is minus ten *because* sixteen plus minus ten is six.

Counting up, meet counting down.

The need to see in **AEM** some way of undoing what addition does has been met, a broken symmetry unbroken.

The natural numbers, the negative numbers, and zero collectively make up the system of the integers. They are a *system* because they are coordinated and because the laws of their behavior are properly derived from their nature.

What remains to be assessed is the way in which **AEM** endows the negative numbers with their identity.

A NEGATIVE IDENTITY

How, to begin with, are the negative numbers generated?

The negative numbers $-1, -2, -3, \ldots$, arise when the natural numbers $1, 2, 3, \ldots$, are subtracted from 0. The negative number $-x$ is just the result of subtracting x from zero, or $0 - x$. But $0 - x$ yields, for every natural number $1, 2, 3, \ldots$, a corresponding negative number: $0 - 1$ equals -1, and $0 - 2 = -2$, a separate clause serving to create a negative number for each natural number. But the formula $0 - 1$ itself follows from the equation $x + z = y$ when the number one is subtracted from zero. The read-off can be read right off: $1 + z = 0$, and therefore there is some number z that makes this equation true. It cannot be any positive number, since one plus any positive number is greater than zero. What is left must be otherwise, and, by convention, it is called a negative number.

What is the *order* of the negative numbers? This is a second question.

The positive numbers $1, 2, 3, \ldots$, are ordered by their size. One number is less than the next one up and greater than the last one down.

The elementary properties of order among the integers (numbers positive *and* negative) are derived entirely from properties of order among the positive numbers. One of them is less than another if their difference is positive. The number five is less than the number eight because eight minus five is three, and three *is* positive. And, more generally, for every number x and y, that x is less than y *means* there is some number z such that $x + z = y$.

But, given the universal scope now assigned to the equation $x + z = y$, all that is needed to impose order on *all* of the integers is the order already imposed on *only* the positive integers. This represents an astonishing degree of leverage.

Whereupon the order of the negative numbers. They go down: they are getting smaller.

Minus one is less than zero.

Does this claim require justification? But look here. Minus one is not equal to 0, nor is it equal to any positive number. Otherwise, it would not be *minus* one. If so, it must be less than zero.

The same conclusion flows directly from the relevant definitions themselves:

Primo: the number -1 equals the number $0 - 1$ *by the definition of the negative numbers.*

Secondo: The number $0 - 1$ equals some number z such that $-1 + z = 0$ *by the definition of subtraction.*

Terzo: But if $-1 + z = 0$, then -1 is less than 0 *by the definition of order.*

The first of these declarations identifies the negative numbers as the residue left when something is taken away from zero; the second identifies the residue as precisely the numbers needed to satisfy an equation; and the third derives from the first and second declarations a condition of order imposed on the negative numbers.

Take away something, see what it is, see how it satisfies.

There remains the matter of just how the negative numbers behave under the operations of addition and multiplication. It is with this question that an almost farcical degree of inventiveness—debts, distances, even divorces and deaths—is employed to assure nervous and uncomprehending students that, while minus five plus minus five is minus ten, minus five times minus five is plus twenty-five.

Nor are students the only ones baffled. How is it, Sadi Carnot asked in the early nineteenth century, that, while minus three is *less* than two, minus three *times* minus three is *greater* than two squared?

What debt handles *that*?

The behavior of the negative numbers under addition and multiplication is derived from first principles—from *mathematical* first principles. The real world, in which debts are made, distances covered, divorces consummated, and death holds dominion over all, has nothing to do with it.

For the first time in the elaboration *of* **AEM**, everything is internal *to* **AEM**.

As much as the invention of the calculus, the incorporation of the negative numbers into **AEM** divides the history of thought.

Before are the long centuries of trial and error in which mathematical ideas were drawn from and directed toward experience.

And after, the period from roughly the end of the eighteenth century to the present, when these mathematical ideas were grasped in all of their chilly remoteness from anything beyond themselves.

20

Certain mathematicians have it—the Tingle, I mean.
They sense what's coming.

THE TINGLE

It is in a treatise entitled *Trigonometry and Double Algebra*, and published in 1849, that Augustus De Morgan argued that algebra "is the science of symbols and their laws of combination." The language is new and it is provocative. The *science* of symbols? The *grammar* of symbols—of course. But symbols are arbitrary because man-made. How could they be the subject of a science? The symbols used in algebra, De Morgan went on to write, are unusual in that they have no sense: "No word nor sign of arithmetic or algebra has one atom of meaning." This is true of their synonyms in ordinary language. "In abandoning the meaning of symbols," De Morgan writes, "we also abandon those of the words which describe them." *Addition* is thus "a sound void of sense."

De Morgan had gotten the Tingle, and his ideas, although poorly expressed, came to light and then to life a half-century later. In the early years of the twentieth century, David Hilbert attempted to reduce mathematics in all of its richness to the *symbols* by which it is expressed. Hilbert had been dismayed by the paradoxes of set theory. No man enjoys being ejected from paradise. In drawing up what widely became known as the Hilbert Program, Hilbert's aims were prophylactic. Mathematicians may think and write as if they had access to sets or groups or numbers, but what they can see or seize are their symbols. *They* are concrete; *they* are in plain sight; *they* can be controlled. Controlling them is the first step in gaining control over the world to which they refer. It is a step, Hilbert argued, the mathematician best accomplishes by coolly rejecting the meaning of his symbols in favor of their syntax, a matter of how symbols are put together and what rules they obey.

ALGEBRA

The Tingle took hold of De Morgan in another way. "If anyone were to assert," he writes, "that + and – might mean reward and punishment, and A, B, C, etc., might stand for virtues and vices, the reader might believe him or contradict him, but not out of this chapter [i.e., book]." The first Tingle had De Morgan deciding that the "words or signs of arithmetic" were "void of sense"; but the second pushed him in the opposite direction. Far from being meaningless, the words and signs of algebra have any number of possible interpretations. They are polyvalent.

Did De Morgan listen to what he had heard? A little bit. The Tingle did what Tingles do. It excited his curiosity.

The work itself other mathematicians would work out.

OLD

Algebra is an old subject. Cuneiform texts from the Babylonian era discuss word problems in terms that have not changed very much in four thousand years.

"I found a stone but did not weigh it, and after I weighed out six times its weight, I added two gin and added one-third of one-seventh multiplied by twenty-four."

What, the scribe then asks, was the original weight of the stone?

It is moving to think of a line of scribes, Sumerians to begin with, or perhaps Chinese, and then Babylonians, Greeks, Romans, Indians, Arabs, Italians, a multicultural multitude, searching, all of them, for some unknown.

Yet none of this is modern algebra. It is not modern, most obviously, even if it remains current, and it does not share the preoccupations of what is new.

NEW

In the introduction to their text, *Algebra*, Birkhoff and Mac Lane attempt to capture what is modern about modern algebra.

Its immense generality, for one thing. Not "the manipulation of *numbers* . . . but elements of *any* sort."

. . . If anyone were to assert that + and – might mean reward and punishment, and A, B, C, etc. might stand for virtues and vices . . .

And the curious way its structures gain purchase on existence by means of their descriptions, for a second thing.

. . . the science of symbols and their laws of combination . . .

This new way of thinking is brilliantly exemplified by the application of modern algebra to **AEM**.

The system of the integers comprises the natural numbers, the negative numbers, and zero. Addition, subtraction, and multiplication are defined. They are prepared to do useful work. Were it not for division, which is yet missing, the system of the integers would have all of the modern conveniences, and, *yes*, it *is* curious and suggestive that the system of the integers without division makes for a natural intellectual object.

Not very much in the system is elegant, because an impression prevails throughout of axioms, definitions, proofs, and laws that have accumulated over the centuries with no strong driving organization. *Just look back.* The natural numbers are given by God. The number zero does what nothing does. The negative numbers are occupied in confounding every honest intuition. Proofs proceed by induction, and induction proceeds by faith. Addition and multiplication are governed by definitions, thick on the ground as thieves in the night.

Thicker, maybe.

If there is no place in this in which a skeptic might say, *See here, something is not quite right,* or even that something is rotten, *God forbid,* this is because **AEM**, although used by merchants, has at every single stage been purified by mathematicians, displaying precisely the characteristics one might expect of a subject maturing through centuries, but maturing by means of the imperatives of a high mathematical culture.

It is only now, within living memory, that *we* can see that **AEM** has an integrity in its structure, a way of being in the world that in the end does *not* reflect the contingencies of commercial life and so does not reflect any contingencies at all—debts, distances, the devil's hand itself, dismissed and now forgotten.

MEANS OF ASCENT

Modern algebra studies structures such as groups, semi-groups, monoids, rings, ideals, modules, vector spaces, semi-lattices, categories, fields—a long list, but one that is *finite*.

The great mathematicians of antiquity were men of remarkable intellectual power, but, like the Indian mathematicians who were their equals, they did not think in terms of abstract algebra. They required the uncertain alchemy of insight and inspiration to see what they had not noticed; and they did not have it: they lacked the means of ascent.

Two thousand years after Euclid laid down his pen, groups became the first algebraic structure to be studied self-consciously; group theory itself came to maturity in work that Évariste Galois composed on the night before his fatal duel. The idea of a ring—*that* required almost another century to become a part of the mathematician's consciousness, and its name a part of his vocabulary.

The landscape revealed by modern algebra is singular in its topography, which resembles an archipelago in which islands are isolated from one another but stable in their identity. In his brilliant little book, *Basic Notions of Algebra*, the Russian mathematician I. R. Shafarevich addresses this point, and he advances this image. His is a voice of reassurance, calm and grandfatherly. "What does mathematics study?" he asks. Shafarevich had spent his long career studying algebra; and, like any specialist, he has no time to waste elsewhere. What does *mathematics* study? No, no. What, in fact, does *algebra* study? "It is hardly acceptable," he writes, "to answer 'structures' or 'sets with specified relations'; for among the myriad conceivable structures or sets with specified relations, only a very small discrete subset is of real interest to mathematicians, and the whole point of the question is to understand the special value of this infinitesimal fraction dotted among the amorphous masses."

Shafarevich is not in this passage conveying a secret; but he is concealing a mystery. A *common* mystery. The algebraic archipelago is very much like other archipelagoes. Biology is the study of living systems, but if we think of the countlessly many ways in which the constituents of living systems, whether their proteins or even their underlying atoms, might be jumbled up and put together, then only an "infinitesimal fraction" of them are of the slightest interest because only an infinitesimal fraction of them are *alive*. In biology, as in modern algebra, "only a very small discrete subset is of real interest . . . , and the whole point of the question is to understand the special value of this infinitesimal fraction dotted among the amorphous masses."

Why such conceptual stability, whether in biology *or* in modern algebra?

The question can only be addressed by allusion. It cannot be answered. "Animals," Jorge Luis Borges once wrote, "are divided into (a) those that belong to the Emperor, (b) embalmed ones, (c) those that are trained, (d) suckling pigs, (e) mermaids, (f) fabulous ones, (g) stray dogs, (h) those that are included in this classification, (i) those that tremble as if they were mad, (j) innumerable ones, (k) those drawn with a very fine camel's hair brush, (l) others, (m) those that have just broken a flower vase, (n) those that resemble flies from a distance."

Animals are not divided this way, of course, but they *could* be. Much the same thing is true of the algebraic archipelago. It could be otherwise.

And yet it is not.

DER NÖTHER

No one ever remarked that Emmy Nöther was handsome. An early photograph in sepia and muted brown shows a face shaped like a lozenge, brown hair swept back from a high forehead, thick eyebrows arched like two caterpillars, a nose straight across the bridge but bulbous at its tip, the mouth prim, and even sullen. She is wearing a long-sleeved blouse tied at the throat with an oversized black bow, and a high-waisted peasant skirt. It is the face of a woman who very shortly would allow the inhibitions of daintiness to give way. Like everything about this remarkable woman's life, these biographical details are bivalent: if she dribbled her food onto her dresses with little concern, and dressed herself with even less, this is because she was consumed by the endless and passionate mathematical discussion that was her life.

Emmy Nöther was born in Erlangen in 1882, and at her death in 1935—sudden surgery, a mistake of some sort, *oops*—she was widely regarded as the most important woman in the history of mathematics. The praise is due to Albert Einstein. Her friend the Russian topologist Pavel Alexandrov referred to her as *der Nöther*, evidently persuaded that to call a woman a man was a considerable compliment.*

Her particular genius in the twentieth century was to consummate the Tingle that began in the nineteenth. She took her

* There are *three* genders in German: masculine (*der*), feminine (*die*), and neuter (*das*).

degree in mathematics at the University of Erlangen in 1907, her work supervised by Paul Gordan, the King of Invariants. Gordan's research was devoted to a part of mathematics then thought of immense importance in which one and the same thing is, by a process of devilishly difficult enumeration, counted in any number of ways. David Hilbert's first great contribution had been to invariant theory, and his contribution was great because, in proving his basis theorem, he emptied the theory of its relevance. It was Hilbert who recognized Nöther's genius, and, while they were both working with Felix Klein, it was Hilbert who schemed to get her to Göttingen, the greatest of great European mathematical centers, with powerful mathematicians in residence, men who, like mastodons, rumbled when they walked and talked. On learning that a woman, of all things, had by the force of her genius obtained space in the mathematics faculty, members of the philosophy faculty, their jowls quivering, objected. Their solemn objurgations forced Nöther to deliver her lectures under Hilbert's name, a deception that must have pleased the old man, for Nöther was bursting with remarkable new ideas, and it was not the worst thing in the world if no one could quite tell where hers began and his ended.

Emmy Nöther was a pure mathematician, but one, oddly enough, whose influence on physics was almost as great as her influence in mathematics. In 1920, she produced an absolutely extraordinary theorem, among half a dozen named after her, in which she demonstrated that the great conservation laws of mathematical physics (mass, energy, angular momentum) were associated with the symmetries of an underlying mathematical system. Imagine a great fish undulating through the ocean waters. Two motions are involved: forward, and from side to side. If the fish's undulations cancel one another, a swish to the left balanced precisely by a swish to the right, Nöther argued, some physical property is conserved—energy in this example, mass in another. A quiver went up and down the physicists' collective spine. It is a quiver quivering still. Symmetry is one idea—some things do not change; that fish is undulating, but its undulations cancel. Conservation is another idea—physical properties are preserved under change. In Nöther's theorem, two ideas are made one.

Having made her mark in mathematical physics, Nöther made a much greater mark in mathematics. In 1921, she published a paper entitled *Idealtheorie in Ringbereichen (Theory of Ideals in Ring)*; it was this paper more than any other that promoted the Tingle seen intermittently throughout the nineteenth century to its conclusion. The great nineteenth-century mathematicians had seen the group for what it was, and Dedekind, the ring for what it could be. With an easy, confident mastery, Emmy Nöther promoted the ring to the position it now occupies: long hidden and never seen clearly even when it *was* seen before, the ring is the abstract object that encompasses the integers. Her work passed almost at once into the textbooks. Much influenced by Nöther's lectures, and by lectures given by the immensely seductive German mathematician Emil Artin (leather jacket, arctic blue eyes, a frigid stare), the young Dutch mathematician Bartel Leendert van der Waerden caught the brightness falling from the air and in 1930 immured it into a brilliant textbook, one entitled *Moderne Algebra (Modern Algebra)*.

With the ascension to power of the Nazis in 1933, Emmy Nöther's position at the university became as imperiled as her life. She had been born to a Jewish family, her genius itself widely understood by Nazi scum to be the ineradicable symptom of racial defilement.

She was able to flee Germany as a part of a remarkable outflowing, a mournful and desperately unhappy exodus.

And she was able to find employment at Bryn Mawr College. Accustomed to the raffish and bohemian life of a European intellectual, what she might have made of the young women she encountered there, all saddle shoes and books apprehensively clutched to their respective bosoms, I do not know. It is hard to say.

She did from time to time visit the Institute for Advanced Study at Princeton University, where she met former friends, lectured, and discovered, no doubt to her private satisfaction, that at Princeton she was as unwelcome as a woman as in Germany she had been unwelcome as a Jew.

It was Emmy Nöther's particular misfortune — *leider* — to have gone as unremarked and as unaccommodated in memory as she was in life.

21

*Groups play a role in **AEM**; but it is the ring that commands its attention and so incurs its devotion.*

TO THE RINGS, THEIR HEARTS

And for every good reason. Within **AEM**, the ring is the place where cases collect and principles are made plain. Access to the rings is controlled by the clauses in a definition. A triple abstraction is at work, a kind of painful ascent. The first demands of the mathematician (and the reader) a willingness to accept the compression that any set of axioms enforces. With the axioms dominant, theorems of the system are diminished in scale, the tidy village reduced to a twinkling toy far below. The second demands of the mathematician (and the reader: you too) a willingness to see beyond the subject that prompted an acceptance of the axioms in the first place. The integers are rings: true enough; but there are rings beyond the integers, another range of alpine peaks far impossible ahead. And there is finally the curious demand that an axiomatic system imposes, a willingness to see *in* the axioms something created entirely *by* the axioms.

This kind of triple abstraction is by no means strictly mathematical. It takes place in the law. "A contract," Williston intones in his magisterial *Contracts*, "is a promise, or set of promises, for breach of which the law gives a remedy, or the performance of which the law in some way recognizes as a duty." This is bare bones. It corresponds to the first abstraction.

The second abstraction takes place when judges (and sometimes juries) come to see in some bilateral agreement the properties of a promise that make it into a contract. Or not.

In *Raffles v. Wichelhaus,* the Seller in Bombay agreed to sell a certain agreed-upon amount of cotton to a Buyer in London. Shipment was to be made on the ship *Peerless.* As it happened — *but who knew?* — there were *two* ships named *Peerless,* one due to sail in October, the other in December. The Buyer in London expected shipment to be made on the *Peerless* leaving port in October; the Seller in Bombay expected to make shipment on the *Peerless* leaving port in December. When the goods failed to appear in London six weeks after the *Peerless* left Bombay — *but which Peerless?* — the Buyer filed suit, appealing, of course, to the sanctity of the contract *he* had signed. The Seller appealed to the sanctity of the very same contract, and inasmuch as he had already been paid for his cotton, pronounced himself well satisfied.

Sitting in judgment, the court (*Raffles v. Wichelhaus* 2 H. & C. 906 [Ex. 1864]) decided that both men were wrong. A mutual mistake had been made. "The moment it appears," Justice Mellish affirmed, "that two ships called the Peerless were about to sail from Bombay there is a latent ambiguity, and *parol* [i.e., verbal] evidence may be given for the purpose of showing that the defendant meant one Peerless and the plaintiff another. That being so, there was no *consensus ad item,* and therefore no binding contract."

Who knew indeed?

Buyer and seller in arguing over the *Peerless* were addressing just this point. An agreement had been reached, *yes.* But a contract? *No,* as the courts decided.

And, finally, there is the abstraction of final ascent, a willingness to see that a contract is whatever the law of contracts says it is. This willingness came hard. It is not easy to see. In the eighteenth and nineteenth centuries, jurists sometimes wrote as if a contract were a "meeting of minds," the written contract simply a reflection of that meeting. At other times in the nineteenth century, jurists wondered whether a contract designated nothing more than the set of all contracts, a collection. These stratagems now seem anachronistic. A contract is whatever satisfies the laws of contracts, and whatever satisfies those laws is a contract. Beyond this reciprocating exchange, there is nothing.

THE RING

Worked out in stages over the course of more than half a century, the definition of a ring is the work of any number of gifted mathematicians: De Morgan, still tingling after all these years; Richard Dedekind, who introduced the concept formally; David Hilbert, who gave rings their name; and Emmy Nöther, who most of all endowed ring theory with its depth.

A ring, then, is composed of a set of elements.

... *elements of any sort* ...

Which contains two elements 0 and 1.

... *might stand for virtues or vices* ...

In which there are two operations of addition and multiplication.

... *might mean reward and punishment* ...

Such that its properties are completely described by its axioms.

The axioms for a ring include six clauses. Not one of them introduces a new or difficult idea. But no link remains to the natural numbers, to zero of old, or to the negative numbers. *They* are doing what they have always done; but as far as *we* are now concerned, the weight of their really real reality is going to be displaced onto the rings.

.   .   .

THE CLAUSES

1     $0 \neq 1$.
2     $x + y = y + x$, and $x \cdot y = y \cdot x$.
3     $(x + y) + z = x + (y + z)$ and $(x \cdot y) \cdot z = x \cdot (y \cdot z)$.
4     $x + 0 = x$ and $x \cdot 1 = x$.
5     $x \cdot (y + z) = x \cdot y + x \cdot z$.
6   For any two elements x and y, there is an element u such that $x + u = y$.

THE TRANSLATION

1) The elements 0 and 1 are distinct.
2) Addition and multiplication are commutative: they go both ways.
3) And associative.
4) The elements 0 and 1 are identity elements. Whatever the element x, 0 plus x is always x, and 1 times x is always x.
5) Multiplication distributes over addition.
6) Subtraction? Sure.

AN ACCUMULATION OF AFFIRMATIONS

If the integers—positive, negative, and zero—satisfy the axioms for a ring, they become ringlike in nature.

There is that curious phrase: *satisfy the axioms?*

The expression has something of the same meaning in mathematics as the expression "meets the conditions" in the law. When does a promise become legally binding? When it meets the conditions for a contract. In both cases, there is an element of the arbitrary in the decision. If this were not so in the law, judges would have little to do, and neither would the mathematicians.

Do the integers satisfy the axioms for a ring?
Apparently so.
0 and 1 distinct? ✓

The addition of numbers commutative? ✓

And associative? ✓

Properly distributive? ✓

Zero and one as identity elements? ✓

Subtraction well and truly defined? ✓

All the clauses having been checked off, the integers, mathematicians say, have checked out.

But checking out the integers is conditional.

The distributive law has, for example, been checked off, but check-offs are not proofs. They simply record assent.

The fact that the distributive law can be demonstrated for the *positive* numbers is of no account. We are talking now of the integers.

"Right," said one of my students emphatically, as if I were expressing a point he had long found troubling.

QUICK SKETCH, SOME DETAILS MISSING

In the idea of a ring, there is much that suggests the integers as we know them. The rings are nonetheless inadequate to trigger anyone's sense that their clauses have uniquely identified the integers, and so brought into perfect alignment intuition, common sense, and mathematical assuredness. The agenda encompassed by the rings does not quite encompass the cancellation law. The cancellation law for addition is fine. It is law legal in *all* rings.

And so is the trichotomy law, for that matter. But the cancellation law for multiplication is false amidst rings, or, what comes to the same thing, true only of the positive integers. If false, it is needed anyway.

It is needed in high school, as when with a starched toile collar rising from a square gray suit, Mrs. Davis addresses her thoughts to the blackboard. She writes in a clear copperplate hand and speaks as if she had been visited by the spirit of Eleanor Roosevelt. Other teachers scratch the chalk, but Mrs. Davis makes it *clack.*

There on the blackboard is the little polynomial equation $x^2 - x - 6 = 0$, tidy in its symbols and short. It is nonetheless not easily solved by inspection. Some modest technique is needed.

The only thing technical in the technique is the reminder (from high school) that the equation $x^2 - x - 6$ may be expressed as the product of its factors $(x + 2)(x - 3)$.

And since $x^2 - x - 6 = 0$, then $(x + 2)(x - 3) = 0$ too.

Solutions are provided by dividing cases and assuming that either $(x + 2)$ is zero, or that $(x - 3)$ is zero.

If $(x + 2) = 0$, then x must be -2.

And if $(x - 3) = 0$, then x must be 3.

Quite right, impeccably so. It is one or the other and, as it happens, both. But what justifies the assumption that if $ab = 0$, then either $a = 0$ or $b = 0$?

It is no assumption that Mrs. Davis ever made.

Lacking the power to cancel at will, the method of indirect identification would fail, and it would fail most obviously whenever quadratic equations are at issue.

Adapted now to the contingencies of a universe in which zero is inconveniently loitering about, the cancellation law for multiplication affirms that if $ca = cb$ then $a = b$. Is this so, no matter what? Of course not.

It is so only if $c \neq 0$.

Given this concession, it *does* follow that if $ab = 0$ then either a is zero or b is.

For suppose that a is not zero. Then $ab = a0$ is $a \cdot 0$, and a may be canceled, just as the cancellation law suggests. In that case, b must be zero, just as indirect identification demands.

A ring in which the cancellation law for multiplication holds is often called an integral ring, or even an integral domain, but algebra is a subject rich in its nomenclature, and nothing much hinges on the term or terms involved. However rings may be named, the cancellation law for multiplication is in force.

But the definition of a ring is deficient in a second sense. It is too flaccid. It suggests a structure that has no convenient and persuasive system for imposing order on the integers. This may come as a surprise. The integers have *already* been ordered by an appeal to the order of the positive numbers. *They* have been ordered, and the integers have shaped up as a result, but the ordering goes unremarked in anything said about the rings.

Who in particular has said that, among the rings, there is anything like the positive numbers at all?

I never did, and, I assume, neither did you; but the assumption must be made if treating the integers as a ring means treating them in a realistic way.

And there is, finally, the well-ordering principle, no part of the definition of a ring, and thus alien to its nature. But at the same time, useful. Should the definition be enlarged to encompass well-ordering?

I suppose so. It cannot hurt.

To the rings, then, these additions: the cancellation law, the positive integers, and the well-ordering principle.

Welcome all.

22

Nothing about the negative numbers is quite so vexing as the law of signs. Minus two plus minus two is minus four; but minus two times minus two is plus four. In four, the same number turns up in the end, but with a different sign dangling from its nose.

SIGN LANGUAGE

The explanation for the law of signs resides in the place where it might be least expected, and that is the definition of a ring, the dazzle of disclosure emerging once the definition has been given and discussed. The explanation is unexpected because it involves a puzzling contrast: the ring on the one hand, the change in signs on the other. Just how can those otherwise austere axioms handle the alchemy by which negative numbers are made positive in multiplication?

Yet they do.

What requires proof is this: that for the numbers x and y, $(-x)$ times $(-y)$ is equal to x times y. That minus one times minus one is one is a special case.

The proof depends on a logical truism. If A is equal to B *and* it is equal to C, then B is equal to C. This is action at a distance, the equality between B and C forged from the fact that each is equal to A. This is a trifle, because everyone knows it; and a truism, because there is no way in which it could be a falsism.*

If the law of signs depends on a logical truism, it also depends on a contrived formula, a form of words useful precisely to the

---

* The same truism is at work in the proof of the associative law of addition in Chapter 15.

151

extent that under algebraic manipulation, it turns out to be equal to two things not obviously related. The formula is *contrived* because standing alone it reveals neither where it is going nor why it is there. Nonetheless, the proof to come demonstrates that the formula is equal to xy and equal again to $-x-y$. From this, it follows that xy and $-x-y$ are equal too.

Here, then, is the proof, step by (grisly) step.

Consider first the triple sum

$$[xy + x(-y)] + (-x)(-y).$$

Notice that this sum has a familiar form

$$(a + b) + c.$$

Recall the burden of association. In adding things together, it is either the first plus the second, and then the third, or the second and the third, plus the first.

Apply the law of association so that

$$[xy + x(-y)] + (-x)(-y)$$

pops up as

$$xy + [x\,(-y) + (-x)(-y)]$$

when square brackets are shifted to the right.

Go ahead and substitute specific numbers for x and y in order to assure yourself that all is well.

Now apply the distributive law to

$$xy + [x\,(-y) + (-x)(-y)]$$

in order to derive

$$xy + [x + (-x)](-y),$$

with $(-y)$ exerting multiplicative action at a distance on $[x + (-x)]$.
But would you look at this:

$$x + -x = 0.$$

And this:

$$0\,(-y) = 0$$

too.

Thank God for zero.

If

$$[x + (-x)](-y)$$

is just zero,

It follows that

$$xy + [x + (-x)](-y) = xy.$$

Connect the head of this argument to its tail:

$$[xy + x(-y)] + (-x)(-y) = xy.$$

Observe with satisfaction that one half of the proof is complete.

The same reasoning works very nicely in going *from* $[xy + x(-y)] + (-x)(-y)$ *to* $-x\,-y$.

So apply the law of distribution to $[xy + x(-y)]$, getting

$$x[y + (-y)] + (-x)(-y).$$

Cancellation to zero is next, and then right away

$$x[y + (-y)] + (-x)(-y) = -x-y.$$

Head to tail a second time

$$[xy + x(-y)] + (-x)(-y) = -x-y.$$

Whereupon (no magic now)

$$xy = -x-y.$$

A difficult argument? No doubt it is off-putting. So many symbols! Mathematicians nonetheless take it in stride. *Elementary*, they say. But against this view there is the odd fact that the derivation of the law of signs from the definition of a ring was not widely understood until the fifth decade of the twentieth century, when Garrett Birkhoff and Saunders Mac Lane published A *Survey of Modern Algebra*. Their proof had without doubt been a part of common mathematical culture before that. But they made it public and they brought it attention.

The derivation of the law of signs from the definition of a ring is entirely modern. Something very abstract is given—a ring—and then, step by step, a series of incidental identities is used to derive a conclusion entirely hidden within the folds and fabric of the definition. Once those incidental identities are seen, they seem almost obvious.

But neither is the proof satisfying, and if it commands assent, the assent it commands is grudging, so that many students, following the steps in their detail, say, *I can see that*—there is a mini-assent, a giving way—and *I can see that*—another mini-assent, another giving way—*but what I cannot see is why a minus times a minus is a plus.*

But would that sense of resistance be dissipated were the mathematician to have affirmed that a minus times a minus is a *minus*?

Try it out and see. If minus four times minus four is minus sixteen, then minus four times *plus* four is what? Plainly, it is not plus sixteen: otherwise, the distinction between negative and positive numbers would collapse. But if it is minus sixteen, then what of the distinction between minus four times minus four and minus four times plus four?

Apparently, there is none.

And if that is so, just what is the distinction between minus four *itself* and plus four *itself*?

If no distinction remains, what is the point of the negative numbers, and if there is none, let us be done with them.

This is less a proof, of course, and more an exercise in ridicule, one contingent on the premise that if you don't like things one way, you are surely not going to like them another way.

The proof that none of this is so is thus a matter of reassurance as anything else. A minus times a minus is a plus. The other way around is really no better.

A MOMENTARY ALIGNMENT

Some mathematicians are good at one thing, and some good at another.

The Hungarian mathematicians Paul Erdős and George Pólya were remarkably good at solving problems. So confident was Pólya of his skill that he agreed to take the Cambridge Tripos in mathematics without preparation.

He did very well.

"Nothing to it," he remarked modestly to the very considerable fury of the men who had worked diligently for months to prepare the problems that Pólya solved with embarrassing ease.

Nothing to it indeed.

At times, the hard, gifted problem-solving men are ascendant. There is no keeping them down. At other times, powerful theoreticians rise up through the ranks, and with immense natural authority demand that their colleagues look beyond their problems.

The supreme mathematical theoretician of the second half of the twentieth century was the French mathematician Alexander Grothendieck. He thought in terms of the utmost generality; he was persuaded that if only the right level of abstraction could be reached, solutions to the cracked, knotted, walnut-hard problems would fall to the mathematician's lips like ripe grapes. His power stunned other mathematicians, so much so that René Thom left mathematics for mathematical biology, depressed, as he remarked, "by a sense of Grothendieck's crushing technical superiority," and invigorated by his entirely erroneous conviction that among biologists he would be received as a living god.

In the early 1980s, Grothendieck vacated his mathematical career to live alone in a shepherd's hut in the Pyrenees.

The hard, gifted, practical men expressed their appreciation for his genius, took what he had given them, and went right on being hard, gifted, practical men.

They were relieved.

THE OTHER SIDE

At some time in the late 1920s, John von Neumann gave a talk at Oxford University; his subject was his specialty, operator algebras, but von Neumann was a great mathematical prodigy, his gifts so lavish and so unconstrained that he was widely thought to be alien in his intelligence. During this period of his life, von Neumann was entirely modern, a master of various new algebraic structures, his ascent into abstraction both effortless and natural.

The British mathematician G. H. Hardy was in the audience. Hardy was himself a superb mathematician, and both a mathematical snob and something of an aesthete, an enemy of modern life. A *Mathematician's Apology* is a lovely book, written in that peculiar English tone of wistfulness of which A. E. Housman was the poetical master. In his memoir, Hardy mourns the passing of his youth and the decline of his mathematical gifts; his nostalgia is sad because it reveals that nothing in life had prepared Hardy for life.

G. H. Hardy was professionally a number theorist. The discipline is most demanding, but it is not new, and the problems on which Hardy had worked had been set in the nineteenth century. Their origins were far older. Had Hardy been returned to ancient Greece, he could have made himself understood to Greek mathematicians.

The lecture commenced. Von Neumann spoke Hungarian as his native language, but except for Basque, I suppose, he spoke the other European languages as well, all of them with an atrocious Hungarian accent. He dressed immaculately, even as a young man, in expensive and well-cut bespoke clothing, and his lectures, so listeners reported, confirmed entirely the general impression that this was a mathematician of extraordinary power.

What Hardy thought of the somewhat portly von Neumann, jabbing his finger in the air to make a point, and covering the blackboard in symbols and the English language in a Hungarian haze, I do not know. He did not say.

But about the talk, Hardy did have something to say. "No doubt," he remarked, "the young man was highly intelligent. But his subject, was it mathematics?"

Was it *really* mathematics?

23

The Rhind Papyrus—a sheet of sheepskin—was excavated
illegally by Alexander Henry Rhind in Luxor in 1858, and
thereafter passed from one corrupt hand to another until
finally, in a vigorous burst of imperial corruption, it made
its way to the British Museum.

FROM THE ANCIENT WORLD AND BACK

In mathematical archaeology, as in so many other areas of life, money has no memory. Written in hieratic script, which stands midway between hieroglyphic and demotic forms of writing, a scheme that goes from pictures to scrawls, the papyrus was composed by a scribe named Ahmes, but, as he himself indicates, it was copied from a still older account current in the reign of King Amenemhat III, thus placing the original at the very throat of the second millennium B.C. Farther to the east, the Sumerian empire has just fallen; there are as yet no Greeks in Greece, and Europe is all bog, drifting mist, impenetrable forest, and solemn mastodons. The Egyptians alone stare out at the world with knowing eyes.

Ahmes was plainly an intellectual overseer of some sort; he is writing with some asperity about the inadequacies of other scribes; and his tone is almost conversational. A practical problem is under discussion: "A building ramp is to be constructed 730 cubits long, 55 cubits wide, containing 120 compartments, and filled with reeds and beams; 60 cubits high at its summit, 30 cubits in the middle, with a batter of twice 15 cubits, and its pavement, 5 cubits. The quantity of bricks needed for it is asked of the generals. . . ." Standing in the hot Nile River sun, the generals are, of course, impatient to undertake some warlike activity, or at least convey the impression of directed energy and so of usefulness; it is the scribes who bear the burden of their demands, and with disappointing results.

"The scribes are all asked together, without one of them knowing anything," Ahmes remarks sourly.

He then addresses the scribe responsible for the thirty other scribes.

"They all put their trust in you," he writes, "and [they] say, 'You are the clever scribe, my friend! Decide for us quickly.'"

There follows the administration of a form of psychological pressure as familiar today as it was in the ancient world: "Behold your name is famous. Let none be found in this place to magnify the other thirty. Do not let it be said of you that there are things that you do not know.

"Answer us how many bricks are needed for it."

INDIRECT IDENTIFICATION

How many bricks *are* needed? Something is unknown; and it is a number. An equation is a verbal form by which an unknown variable and the clues to its identity are associated in an identity.

An equation very rarely reveals the identity of its unknowns on its face. That some number $x = 5$ is true enough, but trivial. More interesting is the claim that some number times itself is twenty-five: $x^2 = 25$. Clues are there in plain sight, but as a clue cannot be convicted of a crime, so a clue cannot stand in for a number. Before the number may be identified, the equation must be solved. The reciprocating exchange between what is not yet known and a compilation of clues about its nature—this is a great drama of mathematics.

In any equation, there are unknowns and there are clues. The clues are what count. Everyone quite understands what it is to be unknown. The equations expressed by **AEM** are polynomial, and so therefore are its clues.

The expanding concourse between numerals such as 1, 2, and 3, and a number of backseat variables in x, y, or z, makes possible in expressions such as $14x$, $8x^3$, or $5x^2$ a record of the way in which some elementary operation has been performed on some elementary *something*.

These expressions make up an interracial clan—monomials, as they are called.

Clan members may be added to one another to form bigger clans. And why not? The monomials are clubbable, and in every case kin to ordinary arithmetic expressions, such as two times two or six times ninety-four, the laws of arithmetic governing indifferently known *and* unknown numbers.

This is a daring idea, but not a new one, for mathematicians of the Arab Renaissance understood it very well, and it was understood by the Greeks, and, before them, by the Babylonians. As ideas go, it has gotten around. The addition of monomials changes only their name, with conjoint monomials called polynomials. The polynomial $5x^2 + 3x$ thus expresses the addition of the two monomials $5x^2$ and $3x$, and the polynomial $5x^2 + 3x + 7$, the same conjunction, with the number seven thrown in for good luck, as when a Chinese merchant embellishes his lottery ticket.

All of this may be subordinated to a system of classification in two parts. A monomial is from the point of view of **AEM** any expression of the form:

$$ax^n$$

and a polynomial, a conjunction of monomials:

$$a_n \cdot x^n + a_{n-1} \cdot x^{n-1} + \ldots + a_1 \cdot x^1 + a \cdot x + a.$$

Polynomials are interesting to the extent that they so clearly designate the operations that they countenance. Multiplication, *check*, addition, *check*, subtraction, *check*, and exponentiation, too. That's a check. But nothing by way of division. These facts are facts of form, because they follow from the way in which polynomials have been defined, but they are also suggestive because of the way in which the polynomials track so faithfully the numbers that are already a part of **AEM**. This is more than coincidence. It represents the heavy, balanced way in which new objects come into mathematical existence.

DOUBLE ENTENDRE

The idea of a polynomial has a latent ambiguity. The polynomials are expressions or linguistic forms, and so a part of the symbolic apparatus by which **AEM** is expressed. The polynomial $5x^2 - 25$ is indeterminate in meaning. Pending some specification of that irresolute variable x, it has a referential form but no referential content. Who knows what number it might designate?

Polynomials achieve specificity and so usefulness when they are embedded in polynomial equations. The equation $5x^2 - 25 = 0$ *says* that two things are equal. One of them is zero. The other must be rationally determined from the clues that the equation provides. It is multiplied by itself. That is one clue. And multiplied again by five. Another clue. From this product twenty-five is subtracted. A third clue. Leaving zero. A fourth.

What, then, is *it*? Inasmuch as Ahmes asked a similar question, no doubt he would have understood this one as well. Language and techniques have in **AEM** changed profoundly over the course of more than four thousand years, but the impulse behind the inquiry remains the same.

The equation $y = x^2 - 7x + 16$ is rather more ambitious than the equation $5x^2 - 25 = 0$. In its form, it is a polynomial equation, because $x^2 - 7x + 16$ is unmistakably a polynomial expression. But it offers as it stands two unknowns, one in x, and the other in y. The second unknown depends on the first, plainly; but knowing this represents progress without traction. The numbers one and ten serve to make the equation true, and so do the numbers two and six. The equation $y = x^2 - 7x + 16$, although saying that two things are identical, lapses in specificity by leaving out *which* things are equal to which.

The unresolved character of one equation may sometimes be relieved by the addition of a second: in this case, the equation: $y = x$.

Two equations are now doing the work of one. They are $y = x^2 - 7x + 16$ *and* $x = y$.

On substituting x for y, there is $x = x^2 - 7x + 16$, one equation now doing the work of two.

With two equations covering two unknowns, both converge on a single number as their solution. Whereupon there is, in $x = 4$, the resolution of an enigmatic identity, the number four uniquely satisfying the equation and the mathematician's confidence in indirect identification.

Clues have done all that clues can do.

COMMUNITY OF INTERESTS

Functions have in **AEM** a prior existence, both as ways in which numbers are conveyed to numbers, and from a far more austere point of view, as sets of ordered pairs. Addition, multiplication, and subtraction are functions of old, ones taking two numbers to a third. In what follows, the expression $f(x)$ designates a generic function, the symbol f leaning heavily on the variable x (its argument) in order to disgorge a new symbol, and so name a new number in turn (its value).

A polynomial function is *any* function of the form $f(x) = a_{n-1} \cdot x^{n-1} + \ldots + a_1 \cdot x^1 + a \cdot x + a$.

The function $f(x) = x^2 + 1$ thus follows the expanding friendship between x and $f(x)$. At $x = 0$, $f(0) = 1$; at $x = 1$, $f(1) = 2$; at $x = 2$, $f(2) = 5$; and at $x = 10,000$, $f(10,000)$, well, something big. These numbers, coming in pairs, comprising an endless list, one stretching from one side of the infinite, in the negative numbers, to the other, in those that are positive.

There is a community of interest between polynomial functions and the equations that are the basis for indirect identification, one expressed by the function $f(x) = x^2 - 7x + 16$ and the equation $y = x^2 - 7x + 16$. Absent the information that $x = y$, the equation $y = x^2 - 7x + 16$ does just what a function does: it tracks the relationship between numbers in x and numbers in y. The substitution of specific numbers for x determines in a machinelike way the value of y. This is the work of functions, so that $f(x) = x^2 - 7x + 16$ and $y = x^2 - 7x + 16$ say the same thing. A triple identity is at work in $f(x) = y = x^2 - 7x + 16$.

What better community of interests could there be?

If polynomial equations and polynomial functions are alike when they are gathered together in order to specify relations between numbers, they are alike again when clues amount sufficiently to change them into statements that specify some very particular number, the relationships forgotten.

A convenient way in which to express these gathering dependencies is to set everything to zero. If $x = x^2 - 7x + 16$, then $x^2 - 7x + 16 - x = 0$. The equations are identical. They say the same thing. They have the same solutions.

But then, considering that $f(x) = x = x^2 - 7x + 16$, a conversion to zero yields $f(x) = x^2 - 7x + 16 - x = 0$.

The number c such that $f(c) = 0$ is called the *zero* or the *root* of a polynomial function.

The *zero* of a polynomial function is the *solution* of its underlying equation: in this case, $f(4) = 4 = x^2 - 7x + 16 = 4$.

The *method* of indirect identification—what does *it* require?
Solve the equation or the equations.
Or
Determine the zero of its function.
Or
Find its root.

THE RING OF POLYNOMIALS

Bedded down amidst a crown of thorns, the polynomials have a second and far more important identity. They form a ring, and, ringlike, they are liberated from their thorns.

The definition of a ring now makes a second appearance. The first clause, remember, is meant to hold off the nightmare of a ring, like a Latin American political party controlled by a single element; the second covers commutations; the third, associations; the fourth, the restoration of identities; the fifth, distributions and disbursements; and the last conveys confirmation that, within a ring, subtraction is whole and complete.

Like contracts in law, rings in mathematics collect their cases, so that the rings are general enough to collect the chief cases of interest but not so general as to lose any connection to the real world in which numbers are at work and do what numbers do.

Polynomial addition proceeds just as one might expect. The sum of the polynomial $5x^2 + 3x + 2$ and the polynomial $3x^2 + x + 5$ is the polynomial $8x^2 + 4x + 7$.

Multiplication requires nothing more than the willingness to collect terms and put them in their proper place. The polynomial $(5x^2 + 3x + 2)$ times the polynomial $(3x^2 + x + 5)$ is the polynomial $(5 \cdot 3)x^{(2+2)} + (5 \cdot 1 + 3 \cdot 3)x^{(2+1)} + (5 \cdot 5 + 3 \cdot 1 + 2 \cdot 3)x^{(2+0)} + (3 \cdot 5 + 2 \cdot 1)x^{(1+0)} + (2 \cdot 5)x^0$.

All bramble and notational thorns in their earlier incarnation, the polynomials now emerge as kin to the integers in satisfying the definition of a ring. The addition and multiplication of polynomials, these are taken for granted because they proceed precisely as examples would suggest; like the integers, polynomials when added to and multiplied by polynomials yield polynomials in turn.

Suppose, thus, that $P(x)$ and $Q(x)$ are two polynomials — identity unknown, parts unspecified. If we look toward the definition of a ring with an eye toward assent, a familiar checklist appears:

1      $P \neq Q.$ ✓

2      $P + Q = Q + P$ and $PQ = QP.$ ✓

3      $(P + Q) + R = P + (Q + R)$ and ditto for multiplication.   ✓

4      $P + 0 = P$ and $P \cdot 1 = P.$ ✓

5      $P \cdot (Q + R) = P \cdot Q + P \cdot R.$ ✓

6   Subtraction? Yes.   ✓

The moment in which the mathematician reveals that the polynomials *are* rings is very like those moments in literature or myth when some humbled hunchback peels off his hump to reveal himself a king:

> *Stand up and lift your hand and bless*
> *A man that finds great bitterness*
> *In thinking of his lost renown.*
> *A Roman Caesar is held down*
> *Under this hump.*

THE IMPORTANCE OF IDENTITY

Polynomials appeared in the historical record well before their algebraic identity was disclosed in the twentieth century. The great Gauss understood perfectly well the analogy between the integers and the polynomials, and he saw how best to exploit that analogy by developing the arithmetic of the polynomials in a way that was parallel to the arithmetic of the integers. The idea of a parallel arithmetic is an aspect of nineteenth- and twentieth-century mathematics, and it combines two quite different mathematical impulses: the need defiantly to unite what seem radically different objects (the integers, the polynomials), and this is a most radical impulse, while at the same time making the basis of similarity the most trustworthy and old-fashioned arithmetical operations, the ancient parts of **AEM**, so that polynomials and the integers become kin not in virtue of some thrilling secret vice, but because they share the most ordinary of obvious virtues. They can be added up, and multiplied, and subtracted.

The discovery that the polynomials formed a ring did what such discoveries often do: it endowed their history with a satisfying form of coherence. The polynomials are *like* the integers. The basic operations of **AEM** scale upward to encompass far more complicated constructions than the integers themselves would ever suggest. The basic operations of **AEM** are true of a world in which things are unknown: just as $2 + 3$ is equal to $3 + 2$, $x + y$ is equal to $y + x$, the commutative law of addition powerful enough to govern addition with no fixed sense of precisely *which* numbers it is governing. *All* of them. It does not matter. If algebra began by liberating the laws of **AEM** from an attachment to any particular set of natural numbers, it succeeded in progressively enriching the combinatorial matrix in which unknowns are played against one another by means of addition, multiplication, and subtraction. The unknowns remain unknown in a polynomial equation, but are bound now by a compilation of clues that the equation itself provides.

An immense journey of insight and abstraction thus lies between $5 + 3 = 3 + 5$, and $x + y = y + x$, and $(5x^2 + 3x + 2) + (3x^2 + x + 5) = (3x^2 + x + 5) + (5x^2 + 3x + 2)$.

What gives this journey, which is the work of centuries, its importance is the way in which a purely abstract desire to enlarge the margins of possible expressions in **AEM**—from numbers, to variables, to polynomials—has made it possible, when integers and polynomials are both seen as examples of a ring, to recover the sense that the flight into abstraction is not necessarily away from the familiar but a way instead of returning to it.

The fact that polynomials are rings allows the mathematician to submit them to the elementary operations of **AEM** itself, and this submission represents nothing less than the power to solve polynomial equations, and, by solving them, justify the method of indirect identification.

This is evident in the simplest of cases.

There is a number such that when seven is taken from it the result is twenty-five.

What number is *it*?

The translation into symbols: $x - 7 = 25$.

And its solution:

Subtract 25 from both sides of the equation, $x - 7 - 25 = 0$, and see at once that $x = 32$.

Whence the easy confidence that $25 - 25$ *is* 0.

From what source permission to subtract anything at all from anything else?

And if -7 plus -25 is -32, as it seems to be, how is that step justified?

Or the next step, which reveals at last that $x = 32$? Just how has -32 become positive by crossing the equator at zero?

These are all questions that no Babylonian could answer, but that we, heirs to the centuries, can. In some measure, but not every measure, the method of indirect identification in which equations encompass some unknown works because *polynomials form a ring*.

ALL THE BORDER POSTS

If interested in finding unknowns, ancient mathematicians were also interested in saving time. Equations, they understood, might well be solved by the application of a formula, something all-purpose and general, a mechanical scheme, or an algorithm, to use a new word for an old idea.

Babylonian mathematicians considered the quadratic equation $ax^2 + bx + c = 0$, and they devised a number of partial formulas for its solution. In the seventh century A.D., Brahmagupta offered the world a general formula for the solution of *any* quadratic equation:

$$x = -b \pm (\sqrt{(b^2 - 4ac)})/2a$$

and the world at once determined that formulas of this sort were a very good thing.

Thereafter, mathematicians were mad for formulas, and in the sixteenth century, a number of Italian mathematicians, the fretful Cardano among them, published algorithms by which third- *and* fourth-degree polynomial equations could be solved mechanically. So great was the importance attached to such formulas that mathematicians cheated one another to obtain them.

Refreshed by plagiarism, Italian mathematicians were nonetheless unable to find a formula for the solution of quintic equations.

What they could not do could not be done. In the early part of the nineteenth century, Paolo Ruffini and Niels Abel both demonstrated that there did not exist a formula applying indifferently to *any* equation of the form $ax^5 + bx^4 + cx^3 + dx^2 + ex + f = 0$.

On the night before his death in some frightful duel, and bursting as much with talent as with desire, the twenty-year-old Évariste Galois looked anew at the roots of polynomial equations, and for the first time saw the system of symmetrical constraints that determined which could be solved and which could not.

And another word before I leave this topic, the last.

The very simplest of polynomial functions may well lack roots in **AEM**. The function $f(x) = x^2 + 1$ is an example. There is *no* number c that sends $x^2 + 1$ to zero.

But this is so in **AEM**, I must stress, and it is *not* generally so. In his doctoral dissertation, Gauss provided the first proof of the fundamental theorem of algebra: *every* polynomial function has its roots.

In the case of $f(x) = x^2 + 1$, and so many other functions, their roots lie beyond the number systems encompassed by **AEM**. It was just this sense of frustration with the number systems of **AEM**, an appreciation of their limitations, that drove mathematicians to the construction of the real and the complex numbers.

Results such as this mark the place on the map where **AEM** ends and something new begins. They are like the lights of border posts in Eastern Europe, the place where one country gives way and another begins.

24

No one persuaded that half a loaf is better than none is
apt to scruple at the number one-half.

IN DIVISION, THE LAST OPERATION

Take the half. Use what you need.

What could be simpler?

If familiarity is any guide, nothing *is*. The ancient world understood fractions and understood why they were needed. The Rhind Papyrus is, among other things, the gossip of technicians, an internal record of tough practical men writing for one another and for their apprentices. The papyrus very successfully conveys the urgency of its calculations. Granaries must be divided, bread apportioned, fields placed under cultivation, soldiers assigned their tasks, these practical problems dominated by the imperatives of an agrarian society always close to scarcity. If the mathematicians of the Rhind Papyrus seemed peevish and even irascible to their subordinates, that is because *their* taskmaster was the Nile, as stern and as unforgiving as the lash. For their calculations, ancient mathematicians needed the fractions, a stable sense of part to whole. They did not worry overmuch about what any of it meant. They made use of shortcuts, tools of the trade. They had no time to waste. They were councilors to the state, after all. Theories they left to the Greeks—still unborn, but nevertheless waiting patiently by the wheel of time so that they could show the Egyptians how it was all done.

For the Sumerians, Babylonians, Egyptians, the Greeks, and, hidden by the earth's great hump, the Chinese, the fractions have been a part of **AEM** since the beginning, items in the Inventory, friends almost as intimate as the natural numbers themselves.

PART TO WHOLE

The fractions are the numbers one-half, three-quarters, five-eighths, and numbers like them—like them, that is, in consisting of two natural numbers in a fixed order. *Any* two natural numbers. And in a fixed *order* too. There is an obvious difference between two-thirds and three-halves. The first trades in thirds, the second in halves. It is better to get three halves of what one wants and one half of what one doesn't than the reverse. The traditional notation for the fractions, in which one number is mounted above the other, conveys this gracefully. Logicians, it is true, often write fractions as ordered pairs, ½ appearing as <1, 2>, but this device does little beyond placing on one line what tradition has always placed on two.

Coming as they do in pairs of natural numbers, the fractions are very conveniently imagined as standing for part to whole. *One* loaf, but *two* slices. This very much suggests that the fractions capture something about the operation of division. And surely this is true. One loaf has been *divided* in two. The operation, whether dividing, cutting, slicing, parceling, or even chopping, comes first. The fractions are in their service. How curious, then, that the fractions *themselves* are often explained by means of the very operation they are meant to represent or explain. The fraction one-fourth is, after all, one *divided* by four. If division must be assumed as an operation in order to make sense of a fraction, then a fraction cannot be assumed as a number in order to make sense of division.

Like the natural numbers themselves, the fractions are constrained in the circumstances of their use. A loaf of bread may be divided in two, but not mud—the phrase *half a mud* is without sense. Mud is not divisible until it has been divided into splotches, dabs, stains, or spots. It seems ungenerous to settle the matter of how mud may be divided by appealing to splotching, dabbing, staining, or spotting, for these activities suppose a prior division of mud into its parts: a splotch of mud is some part of the whole.

Further analytical refinement does not seem easily accessible, and from the perspective of **AEM**, perhaps it is prudent to say that some things may be divided and others not.

Mud we might leave to the philosophers. They love the stuff.

ONE FOR TWO

Arising from the need to divide things, the fractions are themselves divided. The number one is one number; but the number one-half is either two numbers in one, or one number in two.

The impression conveyed by the fractions of two-for-one or even two-*in*-one — is it a matter merely of the way in which they are written?

Decimal notation suggests so. Two in one, or one in two? Nothing of the sort. The fraction one-half contains the seeds of its own dissolution and subsequent rebirth, with one-half emerging as the elegant decimal number point-five.

Two numbers have become one.

Decimal notation is a *general* scheme for the representation of the fractions, one elegant in its economy.

To the left of a decimal point, there is an ordinary integer.

To the right, fractions in tens, one hundreds, one thousands, and so on up to any finite power of ten.

A decimal fraction amalgamates the integer and its fractions by adding them together in one complicated formula.

$$Z + a_1 10^{-1} + a_2 10^{-2} + a_3 10^{-3} + \ldots + a_n 10^{-n},$$

where the Z (from the German Z*ahlen*, or numbers) stands in for the integer of interest.

The expression

$$1 + \tfrac{3}{10} + \tfrac{1}{100} + \tfrac{4}{1000}$$

is a decimal fraction, one that corresponds to the ordinary fraction $\tfrac{1314}{1000}$.

A decimal fraction is one thing, a decimal *number* another. To get the number from the fraction, it is necessary only to keep the fraction's numerators while discarding its denominators, the decimal point serving to separate the integer from the fraction that follows. In place of $\tfrac{1314}{1000}$ — rather ungainly, let us be honest — there is 1.314, sleek as a seal, and as easy to train; all that is needed to use it is a willingness to keep track of the decimal point and the places that it commands.

A number expressed in decimal notation thus has the form

$$Z.\, a_1\, a_2\, a_3 \ldots a_n.$$

The decimal fraction $1 + \tfrac{3}{10} + \tfrac{1}{100} + \tfrac{4}{1000}$ is all fraction, the decimal number 1.314 all number, but the decimal fraction and the decimal number designate one and the same number.

Lovely, brisk, efficient, elegant; and entirely free of that bewildering two-for-one business characteristic of the fractions.

In response to these immodest claims, there is only the obvious: a notational scheme that *begins* with the idea of a decimal fraction can hardly be used to suggest that, when it comes to the fractions, we might get rid of them all.

Whether two-faced, two-sided, or two-headed, the fractions are inexpugnable.

Notation has nothing to do with it.

.   .   :

WHAT THE FRACTIONS ARE NOT

Whatever the fractions such as one-half or nine-tenths may be, they are not integers. A scheme that has encompassed the positive numbers, zero, and the negative numbers does not encompass the fractions. This is odd. The fraction one-half is in the number one and the number two, a composite of two integers. The alloy that results is unlike the base metals from which it is made. The fractions have a nature that no integer shares. The number two is what it is; it is entirely unique, and its position in the tower of the natural numbers can be filled by no other number. The fractions are nowhere near as stable. The fraction one-half, after all, is much the same as the fraction two-fourths, and much the same again as the fraction three-sixths, the fractions one-half, two-fourths, three-sixths, and countlessly many like them sharing a single identity. If the fractions have an identity, it is one that is distributed among like-minded fractions, a great gathering, one for each fraction. And this is one way in which to think of the fractions—as *pairs* of integers, yes, certainly, but then as *sets* of pairs, the set of all those fractions equivalent to a given fraction.

If the fractions are not integers in their identity, they are not integers in another respect as well. The positive numbers are isolated, each appearing on the number line as a short, clean peak, indifferent in its upthrust to numbers around it, alone. With the fractions, it is different. They increase as the positive numbers increase, but this is only to say that, like the positive integers, they have the power to get larger without end. And smaller without end too. But this is something that the positive integers cannot do. On reaching one on the downside, the positive integers face the abyss in zero.

If the fractions are doubly infinite, they are infinite in still another sense, one denied the positive numbers altogether: they are teeming. Between any two fractions, there is always a third.

These points of difference between the integers and the fractions may be reduced to a pair of propositions: There is no positive number between 0 and 1, each number, like stout Cortez, silent on a peak in Darien. But between any two fractions there is always a third. *"Two distincts, division none; Number there in love was slain."*

Is there a proof that this is so?

There is. Certainly there is.

ON THE ONE HAND

The space between zero and one is empty. There is no natural number between them.

The numbers 0, 1, 2, 3, . . . , are discrete, incorruptible in their isolation. In order to prove that between zero and one there is nothing at all, strong principles are required. These the mathematician must import. The well-ordering principle is such an importation. An assumption made within set theory, the well-ordering principle says that every subset of the natural numbers that has at least one member has a smallest member.*

By using the well-ordering principle—"to illustrate its force," as Birkhoff and Mac Lane say with menacing effect—it is possible easily to *prove* that there is no number between zero and one. For suppose that in the number x there were a number where intuition suggests there should be nothing. The number x is greater than zero but less than one: $0 < x < 1$, to put the matter into symbols.

The set of all numbers between zero and one has on this assumption x as a member.

That is enough to trigger the well-ordering principle.

Consider it triggered.

It follows that this set has in some number y a smallest member.

---

* In Chapter 15, the well-ordering principle was used to derive the principle of mathematical induction.

Then $0 < y < 1$.

Multiplying this inequality by y: $0 < y^2 < y$.

From this it follows that y^2 is *smaller* than y.

Ah, but y was supposed to be the smallest of the integers between zero and one, the *very* smallest. A contradiction has been reached, and with it a collapse of confidence in the idea that there *is* a natural number between zero and one.

This argument is very short and very punchy. It conveys a strange disturbing force, just as Birkhoff and Mac Lane affirmed.

ON THE OTHER HAND

Between any two fractions, there is a third, and between the first and the third, a fourth, and between the third and the second, a fifth, any two fractions opening to admit other fractions, this process of internal subdivision uncontrollable, and so proceeding without end, the fractions *pullulating* like some fantastic biological froth, multiplying from within.

The proof that this is so brings about a distinctively mathematical synthesis between radical ideas and their ratification in certainty.

There is, to begin with, a definition of order among the fractions. The number one-third is *less* than the number one-half, and the number six-eighths *less* than the number seven-eighths. That the fraction one-third is less than the fraction one-half *means* that two is less than three, if it means anything at all.

This idea, which is so far a matter of meaning and a tacit appeal to any number of examples, admits of a more disciplined interpretation. Symbolization, with its wearying demands — that is first. It must be. The first symbolic matter is the trifle by which a/b is less than c/d is expressed as $a/b < c/d$.

A two-part stipulation or definition is next. Both parts are familiar from high school, both may be converted from a definition to a theorem, and both are based on the solid, old-fashioned cross-producing identity that if $a/b = c/d$ then $ad = bc$. The proof that this is so is only pages away, but for the moment, I am going to assume that it is so without proof. It is the least that you can do as well.

The first part, the vice of a vice versa: if a fraction a/b is less than another fraction c/d, then their cross product ad and bc track the inequality:

$$if\ a/b < c/d\ then\ ad < bc.$$

The second part, the versa to this vice: if the product of the numbers ad is less than the product of the numbers bc, then the fraction a/b is less than the fraction c/d:

$$if\ ad < bc\ then\ a/b < c/d.$$

Although lacking drama on the occasions of their utterance (we mathematicians are like that), these definitions are dramatic in their effect because they bring the fraction to ground in common products. Witness what follows.

The claim: Between any two fractions, there is always a third. If the fraction a/b is less than the fraction c/d, some fraction F must stand between them:

$$If\ a/b < c/d\ then\ a/b < F < c/d.$$

The argument works by construction. If the fraction F is wanted, here is how it is gotten.

Suppose that the fractions a/b and c/d are not equal, and that a/b is less than c/d.

Suppose? If the fractions were *equal*, why on earth would we be wasting our time on them? There could be nothing between them. If c/d were *less* than a/b, what difference would it make?

So

$$a/b < c/d.$$

From the vice of the old vice versa, it follows that

$$ad < bc.$$

Watch F now emerge:
Pick a number m—any number—and multiply both sides of $ad < bc$ by m:

$$mad < mbc.$$

Now add ba **to both sides:**

$$ba + mad < ba + mbc.$$

Can you do that?
Why not?
Then change $ba + mad$ **to**

$$ab + mad$$

by means of the commutative law, acting to reverse ba.
Then revise $ab + mad$ **to**

$$a(b + md)$$

by allowing the distributive law to turn things inside out.
And inside out again, so that $ba + mbc$ becomes

$$b(a + mc).$$

Commutative and distributive laws are now doing what procedural laws so often do, and that is the work of civilization itself.
So

$$a(b + md) < b(a + mc).$$

But the versa of the old vice versa implies that if $a(b + md) <$ $b(a + mc)$ then

$$a/b < b + md/a + mc.$$

After all $a(b + md) < b(a + mc)$ is just like $ad < bc$ but with a little extra padding thrown in.

Which means that the fraction $b + md/a + mc$ is F, and F is larger than the fraction a/b.

Is there anything left out? No, not really. Precisely the same argument going in the versa direction shows that F is also smaller than c/d.

Larger than the first and smaller than the second, it stands between them.

And it does; a series of absolutely elementary algebraic manipulations, involving nothing more than multiplication and division, has ingeniously revealed a new number standing between two old numbers, the dry shuffling of symbols fruitful, with numbers multiplying in their strange abundance.

CONTRA MUNDUM

In *Brideshead Revisited*, there comes a moment in which Sebastian, doomed by his drinking, enlists Charles Ryder so that the two of them might make a temporary stand against the world—*contra mundum*, as Ryder says. The declaration reflects a dawning sense of division and distinctness on the part of both simpering adolescents.

The elaboration of the integers never really provokes this sense. However strange the negative numbers may be, the system of the integers comports companionably with the physical world, the world of the sciences.

It is with the fractions that the mutual overflow of the physical and the mathematical world, a kind of sloshing, is for the first time impaired, the mathematician finding himself *contra mundum*, the objects at *his* disposal unlike anything anyone else cares to examine.

The number one-half is greater than zero but less than one. If it were zero itself, neither of us would have anything, and if one, then each of us would have twice as much as we might reasonably expect.

If cutting a loaf of bread in two leads to a number between zero and one, cutting that loaf still further should lead to another number, one greater than zero but less than one-half.

And so it does.

In this bread-slicing business, two processes are at work. The one occupying the baker is dividing bread; but the one occupying the mathematician is those proliferating fractions.

These processes are striking in the extent to which they very quickly diverge. Like everything else in the natural world, bread cannot be divided beyond a certain point; bakers willing to try (in the name of Science) simply make a mess of things, with crumbs everywhere instead of slices.

But as the bakers give way, they leave behind a discomfiting idea, one not yet encountered in **AEM,** or encountered anywhere beyond **AEM** once encountered there.

A loaf of bread is like any physical object. It can be cut only so many times before there is nothing left to cut because there is nothing left. The world may suggest but it does not encompass the fractions. From the point of view granted a loaf of bread, the space between zero and one is filled with at most a dozen slices.

From the point of view of **AEM**, the space between zero and one is *teeming*.

In every respect, the mathematical universe is richer and more abundant than one that is physical. Either the physical world must be augmented, or the mathematical world reduced, if harmony is to prevail between them.

The demands of bakery may not seem a likely place for the mathematician to become snooty; but the baker's issue reappears in physics itself.

If the mathematician's number line is dense, the number line to which the physicists defer, on the other hand, is not, and, indeed, physicists are quite confident in pointing to precisely the distance at which considerations of density collapse. It is the Planck length of 10^{-35}.

From a recent article:

> The Planck length is the scale at which classical ideas about gravity and space-time cease to be valid, and quantum effects dominate. This is the quantum of length, the smallest measurement of length with any *meaning*. And roughly equal to 1.6×10^{-35} m or about 10^{-20} times the size of a proton. The Planck time is the time it would take a photon traveling at the speed of light to cross a distance equal to the Planck length. This is the quantum of time, the smallest measurement of time that has any *meaning*, and is equal to 10^{-43} seconds. No smaller division of time has any *meaning*.

The Planck length, these words imply, functions as an *absolute*: there is no getting a smaller distance before distance itself goes bad. The Planck length is thus for particle physicists a *zero* point, the zero meant quite literally as nothing, a region of space that, like the Euclidean point, has no parts, no intrinsic extent;

and from which parts, extent, expanse, and distance are all generated.

It is here, and with these remarkable (but widely made) claims, that the mathematician—any mathematician—must draw the line and say *uh-uh*. It may well be that, in the world of matter, division is finite and comes to an end—the world of matter obeys constraints of its own. But as far as the mathematical world goes, there are fractions without limit, and division without end.

What would a particle physicist say if asked what *one-half* of the Planck length might represent?

Nothing at all, I suppose.

What else *could* he say?

25

Addition, multiplication, and subtraction have been urgent creative forces in the life of mathematics.

THE FIELD OF NUMBERS

Zero and the negative numbers have *both* come about in order to satisfy a requirement of symmetry, a way of completing numbers and systems that were seen to be in place but felt to be inadequate.

Perhaps that sense of inadequacy arose initially for commercial reasons; but at the moment when mathematics reached a certain mysterious stage of self-consciousness, bakers and bookkeepers alike gave way. A placeholder in the world where books are kept (or cooked), zero has its *primary* identity in the world of numbers, where, among its other roles, it appears as the universal answer to the mathematical question: Just what *is* $x - x$, whatever the number x?

Having conquered their initial anxiety, the mathematicians came to see what the bankers never could: that zero and the negative numbers $-1, -2, -3, \ldots$, exist to make whole the equation $x + z = y$.

If the negative numbers and zero are in service to subtraction, the fractions owe their allegiance elsewhere. But they serve just the same ends. They make an elementary operation whole. Just as the equation $x + z = y$ may lack a solution among the positive numbers, the equation $xy = z$ may lack one among the integers.

This is widely thought to be intolerable. Is there a number such that three times that number is seven? If so, then the equation $3x = 7$ must have a solution. If so, it is obviously not one of the numbers 1, 2, 3, 4, . . . At three, products are too great, and at two too little, and just as there is no number between zero and one, there is no number between two and three.

There is in the number two a number that comes *close* to satisfying the equation, and by appealing to the idea of a remainder, it is possible to persuade oneself that all is well. The equation $3x = 7$ has in the number two an integral solution, and in the number one a limping remainder.

Although this device is the subject of a famous theorem in antiquity—the Euclidean algorithm—the operation that emerges is really no one's idea of division.

If there is to be division, there must be numbers of division, and so a felt lack in the way things are balanced, an insecurity about the simple equation $xy = z$, in its turn leads to a completion in **AEM**, new numbers where previously there were none, and with new numbers a sense of symmetry restored.

GOD KNOWS WHAT

At the end of any book about mathematics, inferences collect themselves, just as they do in a detective novel. Clues long overlooked shuffle forward. Perspective is gained. Cases are closed.

The concept of a ring has until now represented the integers $-3, -2, -1, 0, 1, 2, 3, . . .$ in something like their largest aspect. The integers are rings by nature, just as prenuptial agreements are contracts by law. The coincidence between the general concept of a ring and the particular and living presence of the integers was never perfect. It never is. Prenuptial agreements are contracts, but not all contracts are prenuptial agreements, and there are rings that are not much like the integers too. The harmony between the rings and the integers is achieved only when rather special conditions are in place. Whatever the rings in general may be, the ring of integers must contain the positive numbers, a division of things into darkness and light. Otherwise, there would be no order anywhere. And the ring of integers must provide for cancellation, the power to strike out common factors. Otherwise there would be no method of indirect identification. And there is, finally, the well-ordering principle, useful in that it allows the mathematician to say what no one doubted in the first place—that there is nothing between the numbers zero and one. A ring encumbered with these constraints is like a prenuptial agreement in that the first is still a ring, and the second still a contract. But they are different too. This has made for a certain confusion of names. The lawyers do better. A prenup is a prenup, sign it or else, and that is the end of the matter. A ring in which the cancellation law holds is often called an integral ring, and sometimes an integral domain, and an integral ring divided into darkness and light, an ordered integral ring, or an ordered integral domain, and an ordered integral domain satisfying the well-ordering principle is called God knows what.

It hardly matters.

The rings have afforded us a vivid (and profound) sense of the integers; they have exposed their nature by exposing, as in an anatomy lesson, the network that runs just below the otherwise smooth surface in which we add, multiply, and subtract the numbers from one another.

But until now, the rings have, on the subject of the most ordinary fractions, remained unhelpful and unavailing. The rings represent the Platonic form of the integers, and the integers are, by themselves, not enough. They fail to correspond to experience, and so they must be augmented to accommodate the fractions. A very similar development took place in the law of contracts, as the idea of a commercial contract was broadened to include a variety of resolutely *un*commercial promises. In mathematics as in law, in a conflict between the fundamentals of experience, in which both the fractions and certain kinds of promises are given, and an abstract structure collecting them, it is the abstract structure that must give way.

The concept of a ring—that has changed, and it must change yet again. The simplest of changes would involve an acknowledgment of the obvious and a command to make sense of it: The fractions exist. Go find some way of saying that this is so. This way of representing an obvious imperative is too coarse to be of comfort. The fractions *of course* exist, and they exist in order to make division possible. What one wishes from the mathematicians is a more civilized sense of their nature, and a more sophisticated sense of *why* they exist.

IDENTITY AND INVERSE

The numbers zero and one have from the very beginning played a certain role in **AEM**. They would be very conspicuous by their absence.

The numbers zero and one are the base of the great tower from which the natural numbers arise—zero to mark the bottom, and one to force the tower up, one step at a time. But both zero and one have a reverberating effect on the number system itself. Zero is an identity element, and in addition, it carries a number to itself: six *plus* zero is six.

But so is the number one: six *times* one is six.

Zero and one are identity *elements* to the trade, but simple *identities* for us.

The existence of identities—their conspicuous existence—in absolutely elementary mathematics prompts a question. If there are identities at work acting to take numbers back to themselves, are there numbers at work acting to take numbers back to their identities? The question divides according to cases. There is first addition. If six plus zero is six, is there some number such that six plus that number equals zero?

And, of course, there is:

Six plus minus six is zero.

What is more, the requisite reversion to zero may be found for any number a just because $a + (-a)$ is zero.

Such are the additive inverses. The inverses are as numerous as the numbers. It is a remarkable fact that every integer has an additive inverse, but, given the clauses in the definition of a ring, hardly a surprising one. To the extent that the integers are rings, they are bound to admit the possibility of subtraction. The equation $x + y = z$ has a solution, always and forever. If this equation has a solution, integers have additive inverses. For anyone uncertain whether thirty-two has an inverse, set $x + y = z$ to zero. There is thus $32 + y = 0$. It follows that there *is* some number that when added to thirty-two yields zero, the number -32, in fact, the inverse of record for the number in question.

That obvious question now: where are the inverses for multiplication?

The fractions now enter elementary mathematics with a role more sharply defined than any assigned to them by the necessities of daily life. They are multiplicative inverses, numbers that for any integer returns the number to the number one in multiplication. Written as inverses, the fractions appear as a^{-1}. Written as fractions, the inverses appear as $1/a$.

There is now a last adjustment to be made to the primordial concept of a ring. The cancellation law? In place. The positive integers? They are in place as well. Well-ordering? I suppose so. And now this. For every number a which is not itself zero, there is an inverse a^{-1} such that $a \cdot a^{-1}$ equals one. An integral ring (or domain) meeting this demand is known often as a field, and just as often as a division ring.

The fractions are the end of it. With the concept of a field, the arch of absolutely elementary mathematics is complete.

· · ·

NOTHING LEFT TO PROVE

Intuition and the facts of ordinary life have given us the positive numbers, zero, the negative numbers, *and* the fractions. There is no disputing their existence. A very refined process of analysis has, with the idea of a field, ended in a complicated general structure, something like one of the cathedrals designed by the Spanish architect Antoni Gaudí, at once massive and ornate. Why not simply take what we know from long experience and let the algebraic structures go?

The question deserves an answer, the more so since in elementary mathematics life would continue with or without the rings, the fields, or anything besides. The answer, it seems to me, is that as fields, parts of the whole structure of elementary mathematics long scattered and separate acquire a common focus. There is elegance of assumption, and a corresponding power to dissolve the chatter of common experience in favor of something more austere and more dignified.

From the point of view of common commercial concerns, the fractions are an untidy lot. They lack a clear identity, what with one-half now appearing as one-half, and sometime later as three-sixths; and the rules for their manipulation, although learned in childhood, are never justified in adulthood. In multiplying fractions by fractions, go ahead and multiply them, top to top, bottom to bottom; but in dividing them, invert the denominator and then go forward in multiplication. This works out well in practice, but just why does it work at all? The discovery that such questions may all be resolved by means of the single assumption that numbers have multiplicative inverses should act as an aesthetic shock of the sort in which much is dissolved by little.

There is the equation $ax = b$, for example, and with it, the numbers by which it is satisfied. The equation may *always* be solved by letting x be $a^{-1}b$. Acting as the equation's pivot, the inverse of a by itself quite suffices to identify the unknown at x. The inverse of 3 is $\frac{1}{3}$. But what is $3(\frac{1}{3})$ times 10 if not 10?

With the single assumption that numbers have inverses, the fractions take on an ancillary identity. No one proposes to get rid of them; but they are no longer urgent as objects.

Precisely the same assumption accounts for the rules by which fractions are manipulated. Known since ancient times, they may now be *derived* from the definition of a field.

Are the fractions a/b and c/d the same just in case the integers ad and bc are the same? The question is not trivial, for what is at issue is the *identity* of the fractions. Everyone understands, of course, that the fractions one-half and five-tenths are the same. But what is at issue is not *whether* this is known but *how*, and that is another matter entirely. It is the identity of the fractions that, after all, enters into the proof that the fractions are dense, and without a secure sense that their commonplace identity is well-founded, the proof would collapse.

What requires demonstration is this:

$$\text{if } a/b = c/d \text{ then } ad = bc.$$

This is purely a hypothetical statement, and in order to demonstrate it as a whole, it suffices to assume its antecedent and from this assumption, derive its consequent.

So assume that

$$a/b = c/d.$$

By the definition of an inverse,

$$a/b = ab^{-1}.$$

And by the same definition

$$c/d = cd^{-1}.$$

Four in one

$$a/b = c/d = ab^{-1} = cd^{-1}.$$

It is from this identity, together with the associative and commutative laws, that the inner nature of the fractions is forged.

First there is

$$ad = a(b^{-1}b)d$$

because by definition $b^{-1}b$ is 1.

What else could it be?

Second change $a(b^{-1}b)d$ into $(ab^{-1})\,bd$, so that

$$ad = (ab^{-1})bd.$$

How did that work? By means of the associative law acting on $a(b^{-1}b)$, and shifting its parentheses to the left to form $(ab^{-1})b$.

Next appeal to the identity $a/b = c/d = ab^{-1} = cd^{-1}$ so that ab^{-1} gives way to cd^{-1}:

$$ad = cd^{-1}bd.$$

Stop and look.

By the commutative law, $cd^{-1}(bd) = cd^{-1}(db)$, so

$$ad = cd^{-1}(db).$$

Now shift parentheses so that $cd^{-1}(db) = c(d^{-1}d)b$.

$$ad = c(d^{-1}d)b.$$

Watch the two numbers $d^{-1}d$ mutually annihilate each other in one, yielding cb, and by the commutative law bc.
Thus

$$ad = bc.$$

All done: Symbolic magic: Magical symbols.

This argument requires no insight, and hardly any intelligence. It is mechanical, an exercise in which the product of two numbers a and d is tracked through various of its identities until its last identity emerges in the product of the numbers b and c.

The proof tells us nothing new. It is not meant to. It is the telling that is new. A property of the fractions has been derived from the definition of a field, *and nothing more.*

This is what is new.

The other notable properties of the fractions may in the same way be derived, and with their derivation, the fractions as autonomous agents disappear, leaving only their field behind.

END OF STORY

But there is no end to the story of **AEM,** and neither is there a beginning. The Peano axioms mark a place, but it is one place among many others, and by itself it can tell us nothing beyond the fact that mathematics and so **AEM** has roots in some unfathomable aspect of the human mind. The definition of a field marks another place, but it is also one place among many others, and by itself it can tell us nothing beyond the fact that mathematics and so **AEM** calls on all the powers of the human mind to create abstractions and to believe in them.

AEM is, like any part of mathematics, a work of art, but it is, like no other part of mathematics, a collective work, one extended over centuries, the labor of merchants, bankers, and accountants as well as mathematicians, at once near and divinely distant, and for most of us that part of the great mathematical meditation that, because it lies so tantalizingly close at hand, suggests as nothing else can the glory that is beyond.

CONCLUSION

In his memoirs of Napoleon's conquest of Egypt, the Duke of Rovigo tells a story of conflict, honor, and the way in which an elegant moral code offered to men of that time and place a way of shaping their lives into a coherent narrative.

His account turns on the shifting and often incoherent dynastic allegiances of the Ottoman empire. In 1768, the Georgian-born Ali Bey led a successful insurrection against Ottoman rule in Egypt. He was, the Duke recounts, "a man possessed of humane feelings, and of natural talents [and] the only Bey whose memory appeared to be cherished by the Egyptians."

Having acquired authority as the leader of an insurrection, Ali Bey proceeded to lose influence as the governor of a country, his assassination in 1778 bringing his power along with his life to an end.

Among his assassins was his rival, Mourad Bey; the affair described dryly by the Duke as "one of the affrays so common amongst those petty tyrants."

But among Ali Bey's beneficiaries, there had been Hassan Bey, for it was Ali Bey who had made Hassan a bey, the title his access to power and a sign of prestige.

On the one side, a desire for revenge; and on the other, the desire to escape it.

Hassan Bey was "a formidable warrior" — the Duke again — but he was defeated in battle by Mourad near Cairo; and thereafter "he was . . . hotly pursued."

The story now reaches its climax. Having failed to prevail against his enemy, Hassan Bey, the Duke recounts, sought refuge in his seraglio, "soliciting asylum from his favourite Sultana." The thought that in fleeing one's enemies one might with confidence turn to one's mistresses is not current in modern military circles. Or anywhere else, for that matter, but "in eastern countries," the Duke writes in admiration, "the laws of hospitality are held sacred."

A number of exciting adventures ensue, the raffish Hassan Bey escaping the seraglio, disguising himself, evading capture, and in the end allying himself improbably with Mourad Bey.

Whether any of this is true, I do not know, and, I suspect, neither did the Duke.

There remains that moment of sanctuary in the seraglio—a cliché, to be sure, the Near East under Western eyes, but the story curious and compelling nevertheless.

Clichés so often are.

In painting his *Harem Scene with Sultan*, Jean-Baptiste van Mour could not have had the Duke of Rovigo's memoirs in mind; he lived and worked before Napoleon's conquest of Egypt. Still, he knew the Orient, he had lived in Constantinople, and, unlike Jean-Léon Gérôme, an artist whose hand also turned naturally to harems and harem scenes, *he* had had access to the Ottoman court and its palaces.

His *Harem Scene with Sultan*, which forms the frontispiece to this book, is an interior, Dutch in its conception, French in its style, an altogether polished piece of work.

The painting depicts a large rectangular room. The low divan at the far wall, the textured carpets, and the floor tiles are either squares or rectangles. The room is plainly of its time and of its place, but it is modern in the way that it combines geometrical austerity and ornamental elegance. Only three shapes are at work throughout—the square, the rectangle, and, in the various harem women and a low table, the cylinder—but these shapes reflect three aspects of a single shape, the rectangular parallelepiped, even the cylinder nothing more than a rolled-up rectangle. If nothing else, van Mour's painting is a study in the economies of form.

There are nineteen women in the room; slim and tall, they are dressed in the Ottoman style. The two women in the center appear to be playing handheld instruments. One woman arches backward; the other, wearing a fez, faces her as she dances.

On the right side of the room, three women are arrayed around the low table, one of them attending to a seated Nubian servant, the harem's eunuch perhaps; one has turned her head to watch them; and one is about to remove a platter from the table.

Half reclining on a rectangular couch, and visually diminished by the picture's scale, the Sultan is at his ease, his knees carelessly splayed. He is dressed in billowing red pantaloons. A woman of the harem is attending him. Isolated in his masculinity, the Sultan is in conversation with another, seated woman, and she is staring at *him*—the warrior chief at ease in a seraglio, *the* seraglio, to impose the Duke of Rovigo's memoirs on this painting, for it is Hassan Bey in red pantaloons, and Mourad Bey's Sultana who is attending him.

If he is at his ease in the seraglio, the Sultan is nonetheless an alien figure, a warrior amidst women. The painting—it is a subtle work—suggests, but never displays, the contrast that these scenes are meant to evoke, between what is urgent and undivided, and what is complicated but refined, the Sultan *undone* within the seraglio, his fierceness compromised by the sloe-eyed women and the elegant food, the sound of zithers, the cushions of silk, perfume in the warm air, the arts of a refined civilization, but the women as much undone as the Sultan, because he *is* the Sultan and *they* are in thrall to *him*, their great art in the service of a contending force that they cannot in themselves create and could not in any case control.

There is a balance, then, between the Sultan and the harem women, the exquisitely reticulated room conveying the poise of a moment in which what is primitive and thus given, and what is civilized and thus made, find one another in contentment.

I am writing about mathematics.

But you knew that. I hope you did.

ACKNOWLEDGMENTS

I am very indebted to my friend Morris Salkoff for his careful reading of my book, and for his many valuable corrections.

And indebted as well to my editor, Edward Kastenmeier, for *his* careful reading of my book, and his many valuable corrections.

INDEX

Abel, Niels Henrik, 59, 168
Abélard, Pierre, 36–8
absolutely elementary mathematics
    (**AEM**), 3, 15, 31, 64, 69
  in ancient times, 9–10
  application of algebra to, 137
  coherence of, 4–5
  counting and, 5, 54
  definitions used in, 83–4
  geometry vs., 56
  hypothetical statements, 32–4
  inductive proofs in, 103–4
  integrity of structure of, 138
  limits of, 169
  polynomial functions without
    roots in, 168–9
  theory of, 7
  see also fractions; natural numbers;
    negative numbers; zero
abundant numbers, 50, 51n
addition, 57, 79, 183
  associative law and, 97, 102,
    116–20, 147
  cancellation law for, 98–9, 148
  commutative law and, 97–8, 147,
    166
  distributive law and, 100–1
  of negative numbers, 133–4
  notation for, 57–8, 62–3

in polynomials, 160
as progressive, 58
subtraction vs., 58, 59, 129, 130–1
succession and, 51–2, 64, 68–9
trichotomy law and, 99–100
as "void of sense," 135
by zero, 58, 65
addition, definition of, 64–9, 70, 102
  as absent for Peano axioms, 52
  associative law and, 120
  counting and, 62, 69, 74
  and definition by descent, 61–2,
    65, 68, 73, 84–5
  difficulty of arriving at, 4, 59
  justification of, 4, 59
  in modern algebra, 137
  in proof of associative law of
    addition, 118
  in real example, 67–9
  succession in, 68–9
  three clauses of, 64–6
  zero in, 65, 68
Ahmes, 158–9, 161
Alexander the Great, 29–30
Alexandria, Egypt, 123
Alexandrov, Pavel, 140
algebra(s), 3, 11, 18, 135–40
  application to **AEM,** 137
  fundamental theorem of, 169

Index

Bedouin, 83
Berlin Academy, 11
Bhaskara, 123–4
Big Bang, 50
binary, 79
biology, 139
Birkhoff, Garrett, 137, 154, 175–6
Book of Restoration and Equalization
    (al-Khwarizmi), 18
Book of the Thousand Nights and
    One Night, The, 17
Boole, George, 63, 91
Borges, Jorge Luis, 139–40
Brahmagupta, 123
Braunschweig, 53
Brideshead Revisited (Waugh), 33, 179
British East India Company, 92
British Museum, 158
Bryn Mawr College, 142–3

Calamari, John D., 33
calculators, 4
calculus, 3, 112, 115
    creation of, 124, 134
    notation of, 93
    use of, 6
Cambridge University, 90, 155
cancellation, law of, 95, 98–9
    for addition, 98–9, 148
    law of signs and, 154
    for multiplication, 99, 149–50
    rings and, 148–9, 185, 187
    zero and, 99
Cantor, Georg, 11, 22, 25, 54
Cardano, Gerolamo, 168
Carnot, Sadi, 134
categories, 138
Catholic Church, 37
Cayley, Arthur, 91–2
certainty, 26–9, 35

chemistry, 53
China
    algebra in, 136
    essentials of **AEM** grasped in, 9
    fractions in, 170–1
    polynomials known in, 160
Church, Alonzo, 30–2
circles, 28
    squaring of, 111
Clifford, William Kingdom, 91
Cloister School at Notre-Dame, 37
Collegium Carolinum, 53
commutative law, 95
    addition and, 97–8, 147, 166
    associative law vs., 98
    fractions and, 190–1
    and infinite density of fractions,
        178–9
    multiplication and, 97–8, 147
    in ring axioms, 147, 148, 164, 165
    as symmetry, 96–7
complex analysis, 31
complex numbers, 169
conjunction, 34
Connes, Alain, 6–7
conservation laws, 141–2
Constantinople, 194
contracts, 164
    conditions for, 147
    hypothetical statements in, 33
    specific examples of, 184, 185
    triple abstraction of, 144–6
Contracts (Calamari and Perillo), 33
Contracts (Williston), 144
Copernican system, 27
counting, 8
    AEM and, 5, 54
    and definition of addition, 62,
        69, 74
    explanation of, 14–15

Index

counting *(continued)*
  limits of, 8–9
  natural numbers generated by,
    8–9, 45–6, 51–2, 54–6
  of sets, 25
  as simplest arithmetic act, 49, 50
creation, 124–5
Croesus, 32
cuneiform texts, 136

Darwin, Charles, 113
debts, 3, 138
  negative numbers and, 126–7, 134
decimal notation, 172–3
Dedekind, Richard
  background of, 53–4
  on counting, 49–50, 55, 69
  literary voice of, 59–60
  Peano axioms and, 44
  ring work of, 142, 146
deductive theorem, 32
deficient numbers, 51, 52n
definition by descent
  for addition, 61–2, 65, 68, 84–5
  as allegedly circular, 61–2
  and existence of function, 86–8
  in exponentiation, 75–6
  for multiplication, 72–3, 84–5
  recursion theorem as proof of, 85,
    87–8, 89, 102
definitions
  of base, 79
  of division, 70
  of exponentiation, 75–6
  of fields, 187, 191
  of negative numbers, 133
  of ordering, 133
  of rings, 146, 148–50, 151, 154,
    164–5
  of subtraction, 70, 133, 137, 148

used in **AEM**, 83–4
  of well-ordering principle, 108
  see also addition, definition of;
    multiplication, definition of
De Morgan, Augustus, 63, 91,
  92–4
  principle of induction expressed
    by, 103
  on signs of algebra, 135, 136
  work on rings by, 146
Descartes, René, 111
*Description of the Marvelous Canon
  of Logarithms*, A (Napier), 77
Dieudonné, Jean, 56
Diophantus, 123
Dirichlet, Peter, 54
disjunction, 34
distances, 3, 134, 138
  as derived number, 54–5
  negative numbers and, 125–6
  Planck length and, 181–2
distributive law, 95, 100–1
  and definition by descent, 73
  and infinite density of fractions,
    178–9
  and law of signs, 152, 153
  in ring axioms, 147, 164, 165
division, 57, 170–1
  associative law and, 97
  definition of, 70
  as finite in physical world,
    181–2
  fractions and, 171–2, 184, 188
  multiplication vs., 59
  polynomials and, 161
  unnatural aspects of, 70
division ring, 187
dominoes, 104–5
double-entry bookkeeping, 127
Doughty, Charles M., 83

Index

education, mathematical, 93
Egypt, 158–9, 170–1, 193–6
Einstein, Albert, 6, 28, 140
Elements (Euclid), 41–3, 55–6, 103
elliptic function, 11
empty set, 24
energy, conservation of, 141–2
England, 90–1
equality, 16–17
equations, 159
   application of formulas to, 167–8
   polynomials *see* polynomial
     equations
   quadratic, 123–4, 168
   quintic, 168
Erdős, Paul, 155
Euclid, 6, 41–3, 44, 83, 123, 138, 181
   axioms created by, 41, 44
   derivation of numbers by, 55–6
   induction in, 103
   infinite number of primes
     demonstrated by, 95
   legacy of, 41–2
   as mental training, 42–3
Euclidean algorithm, 184
even numbers, 50
   as well ordered, 108–9
exponentiation, 71–8, 79
   definition by descent in, 75–6
   definition of, 75–6
   of number two, 86
   in polynomials, 160
   three clauses of, 75–6
   with zero, 76
exponents, bases and, 75

falsity, 33–4
Fermi, Enrico, 43
field(s), 138, 188
   definition of, 187, 191

fixation, 125
formulas, 16–17
   as applied to equations, 167–8
Foundations of Analysis (Landau), 5
fractions, 3, 4, 7, 59, 170–82, 188
   in ancient times, 170–1
   as counterintuitive, 180–1
   decimal notation and, 172–3
   definition of order among, 176
   division and, 171–2, 184, 188
   as infinite, 174
   infinite density of, 176–9, 180–1
   integers vs., 174–5
   multiplication of, 188
   as multiplicative inverses, 187,
     188–90
   rings and, 185–6
Frege, Gottlob, 63
French Academy of Science, 114
Fulbert, Canon, 39–40
function 2^x, 87
functional notation, 57–8
functions, 162–3, 168–9
   and definitional descent, 86–8
   see also specific functions
fundamental particles, 51

Galois, Évariste, 59, 138, 168
Gattopardo, Il (Lampedusa), 48
Gaudí, Antoni, 188
Gauss, Carl Friedrich, 53, 59
geometry, 3, 55–6
   AEM vs., 56
   analytic, 112
   arithmetic's relationship with, 125
   axioms of, 41–3, 44, 83
   use of, 6
Gérôme, Jean-Léon, 194
Gödel, Kurt, 29
Gordan, Paul, 141

Index

Great Dictionary of the Natural
Numbers, 79–82
Greece, 158
algebra in, 136
fractions in, 170–1
logic in, 29–31, 36
polynomials known in, 160
see also Elements (Euclid); Euclid
Grothendieck, Alexander, 156
groups, 138

half-line, 121–2, 124
Hamilton, William Rowan, 91
Hardy, G. H., 156–7
Harem Scene with Sultan (van
Mour), 194–5
Hassan Bey, 193–4, 195–6
Héloïse, 39–40
Herodotus, 32
Herschel, John, 93
Hersh, Reuben, 16–17
Hilbert, David, 135–6
Nöther's genius recognized by,
141
ring work of, 146
on set theory, 22, 135
Hilbert Program, 135–6
Hindus, 18
Historia Calamitatum (Abélard), 38
Housman, A. E., 157
hypothetical statements, 32–4

ideals, 138
Idealtheorie in Ringbereichen
(Nöther), 142
identity elements, 72, 73
in ring axioms, 147, 148, 164,
165
see also one; zero
imperfect numbers, 50

India
algebra in, 136, 138
positional notation in, 18
zero in, 21
indirect identification, 159–61, 162,
163
induction, principle of, 103–5
associative law proved by, 106,
116–20
as axiom, 105
as derived from well-ordering
principle, 108–10, 175n
Ratchet proof of, 105–8
Institute for Advanced Study, 143
integers, 131–2, 137
fractions vs., 174–5
ordering of, 150
polynomials vs., 165, 166
ring axioms satisfied by, 142, 144,
147–50, 165, 166–7, 184–5
integral domain, 150, 185
integral ring, 150, 185
Introduction to Mathematical Logic
(Church), 30–2
*Introduction to Mathematical
Philosophy* (Russell), 12–13
Introduction to Metamathematics
(Kleene), 85
invariant theory, 141
Italy, 136

Jacobi, Carl Gustav, 90
Jerome, Saint, 39
Johnson, Samuel, 117

Kennedy, Hubert, 46
Kepler, Johannes, 26
Khwarizmi, Abu Ja'far Muhammad
ibn Musa al-, 17–18
Kirkman, Thomas Penyngton, 91

Index

Index

operation, 57
 see also specific operations
operator algebras, 156–7
Oppenhcimer, J. Robert, 29
ordered pairs
 fractions as, 171
 sets of, 86–7, 162
ordering, 15
 definition of, 133
ordinary differential equations, 43
Organon (Aristotle), 29
Ottoman Empire, 193–6
Ovid, 27
Oxford University, 90, 93

Pacioli, Luca, 126–7
parameters, 116, 117
parentheses, 66, 67, 77, 130
Paris, 36–8, 43
Pascal, Blaise, 103
Pauli, Wolfgang, 53
Paulos, John, 16*n*–17*n*
Peacock, George, 91, 93
Peano, Giuseppe, 43, 44–7, 63
Peano axioms, 44–6, 49, 191
 addition and, 52, 65
 fifth, 45, 46, 103–4, 108, 109–10,
  117, 120
 principle of mathematical
  induction in, 103–4, 108,
  109–10, 117, 120
 succession in, 45–6, 49, 52, 65,
  109–10, 117
Peirce, C. S., 63
perfect numbers, 50, 51*n*
Perillo, Joseph M., 33
physics, 51, 53–4, 141–2, 181
Planck length, 181–2
Planck time, 181
plane, 55

Plato, 29
Platonic forms, 37–8
points, 55, 121, 181
Pólya, George, 155
polynomial equations, 149, 159–63
 and roots in **AEM,** 168–9
 specificity of, 161–2
polynomials, 159, 160–1, 164–5
 ambiguity of, 161
 integers vs., 165, 166
 ring formed by, 164–7
positional notation, 16, 18–21,
  79–81
positive numbers *see* natural
  numbers
Prausnitzer, Fanny, 11
prenuptial agreements, 184, 185
prime numbers, 51
 infinite number of, 95
 theorems of, 95
Princeton University, 143
probability, 126
procedural mathematics, 95
pronouns, 63
proofs
 complicated structure of, 5–6
 deduction theorem, 32
 of definitions by descent, 85, 87–8,
  89, 102
 of law of signs, 151–5
 outside of math, 28–9
 Ratchet, 105–8
 of recursion theorem, 88
 of trichotomy law, 99
proofs, inductive, 103–4, 116–20,
  137
 associative law and, 106, 116–20
propositional connectives, 34
Ptolemy, Claude, 26–7, 28
Ptolemy's theorem, 28

Index

Index

A NOTE ABOUT THE AUTHOR

David Berlinski was born in New York City. He received a B.A. from Columbia College and a Ph.D. from Princeton University. He lives in Paris.

A NOTE ON THE TYPE

This book was set in Electra, a typeface designed by William Addison Dwiggins (1880–1956) for the Mergenthaler Linotype Company and first made available in 1935. Electra is a simple, readable typeface that attempts to give a feeling of fluidity, power, and speed.

COMPOSED BY

Creative Graphics, Allentown, Pennsylvania

PRINTED AND BOUND BY

Berryville Graphics, Berryville, Virginia

BOOK DESIGN BY

Robert C. Olsson